T0337840

Advances in Nanomaterials and Nanostructures

Advances in Nanomaterials and Nanostructures

Ceramic Transactions, Volume 229

Edited by
Kathy Lu
Navin Manjooran
Miladin Radovic
Eugene Medvedovski
Eugene A. Olevsky
Chris Li
Gurpreet Singh
Nitin Chopra
Gary Pickrell

The American Ceramic Society

WILEY

A John Wiley & Sons, Inc., Publication

Published by John Wiley & Sons, Inc., Hoboken, New Jersey.
Published simultaneously in Canada.

For general information on our other products and services or for technical support, please contact our Customer Care Department within the United States at (800) 762-2974, outside the United States at (317) 572-3993 or fax (317) 572-4002.

Wiley also publishes its books in a variety of electronic formats. Some content that appears in print may not be available in electronic formats. For more information about Wiley products, visit our web site at www.wiley.com.

Library of Congress Cataloging-in-Publication Data is available.

ISBN: 978-1-118-06002-5
ISSN: 1042-1122

oBook ISBN: 978-1-118-14460-2
ePDF ISBN: 978-1-118-14457-2

Printed in the United States of America.

10 9 8 7 6 5 4 3 2 1

Contents

NANOTECHNOLOGY FOR ENERGY, HEALTHCARE AND INDUSTRY

NANOLAMINATED TERNARY CARBIDES

Preface

There have been extraordinary developments in nanomaterials in the past two decades. Nanomaterial processing is one of the key components for this success. This volume is a collection of the papers presented at three nanotechnology related symposia held during the Materials Science and Technology 2010 conference (MS&T'10), October 17-21, 2010 in Houston, Texas. These symposia included Controlled Processing of Nanoparticle-based Materials and Nanostrctured Films; Nanotechnology for Energy, Healthcare, and Industry; and Nanolaminated Ternary Carbides and Nitrides (MAX Phases).

Nanoparticle-based materials and nanostructured films hold great promise to enable a broad range of new applications. This includes high energy conversion efficiency fuel cells, smart materials, high performance sensors, and structural materials under extreme environments. However, many barriers still exist in understanding and controlling the processing of nanoparticle-based materials and nanostructured films. In particular, agglomeration must be controlled in powder synthesis and processing to enable the fabrication of homogeneous green or composite microstructures, and microstructure evolution must be controlled to preserve the size and properties of the nanostructures in the finished materials. Also, novel nanostructure designs are highly needed at all stages of bulk and thin film nanomaterial formation process to enable unique performances, low cost, and green engineering. This volume focuses on three general topics, 1) Processing to preserve and improve nanoscale size, structure, and properties, 2) Novel design and understanding of new nanomaterials, such as new synthesis approaches, templating, and 3D assembly technologies, and 3) Applications of nanoparticle assemblies and composites and thin films.

We would like to thank all symposium participants and session chairs for contributing to these high-quality and well attended symposia. Special thanks also go out to the reviewers who devoted time reviewing the papers included in this volume. The continuous support from The American Ceramic Society is also gratefully acknowledged. This volume reflects the quality, the scope, and the quality of the presentations given and the science described during the conference.

KATHY LU
NAVIN MANJOORAN
MILADIN RADOVIC
EUGENE MEDVEDOVSKI
EUGENE A. OLEVSKY
CHRIS LI
GURPREET SINGH
NITIN CHOPRA
GARY PICKRELL

Controlled Processing of Nanoparticle-Based Materials and Nanostructured Films

EFFECT OF FOCUSED ION BEAM PATTERNING ON ENLARGING ANODIZATION WINDOW AND INTERPORE DISTANCE FOR ORDERED POROUS ANODIC ALUMINA

Bo Chen, Kathy Lu, Zhipeng Tian
Materials Science and Engineering, Virginia Polytechnic Institute and State University
Blacksburg, Virginia 24061, USA

ABSTRACT

Highly ordered porous anodic alumina with alternating-sized pores and in hexagonal and square arrangements has been produced with focused ion beam patterning guided anodization. Deeper focused ion beam patterned concaves induce better developed pores during the anodization. Focused ion beam patterning also effectively enlarges the anodization window; ordered alternating-sized nanopore arrangement and square arrangement with 150 nm interpore distance can be produced at 40-80 V potentials. Under the guidance of FIB patterned concaves in Moiré patterns, different alumina nanopore arrays in Moiré patterns can be obtained after the anodization.

INTRODUCTION

In recent years, nanomaterials have attracted great interest due to their unique electronic, magnetic, and optoelectronic properties and a broad range of applications in new nano-devices.[1-8] Among different fabrication methods of nanomaterials, templating based on porous anodic alumina has the advantage of offering uniform diameter and controlled aspect ratio pores, which can then be used to create high density and perfectly vertical nanorod and nanowire arrays.

Self-organized porous anodic alumina cannot provide highly ordered nanopores across large areas, the honeycomb-like structure is limited in only several micrometer scale. Even though two-step anodization process can increase the area of highly ordered nanopore arrays, the anodization condition is limited in a narrow window: 63 nm (0.3 M sulfuric acid, 25 V),[9-11] 100 nm (0.3 M oxalic acid, 40 V),[12-13] 500 nm (0.3M phosphoric acid, 195 V),[14-15] and only hexagonal arrangement of uniform diameter pores can be produced. Attempts to fabricate porous anodic alumina with arrangement other than hexagon and with different pore diameter have been made. Square arrangement of square nanopores was synthesized using nano-indentation.[16-18] Similarly, triangular anodic alumina nanopores with a graphite lattice structure were synthesized with the guidance of nano-indented graphite lattice structure concaves.[19] With the guidance of the focused ion beam created gradient- and alternating-sized concave patterns, hexagonally ordered gradient- and alternating-sized nanopore arrays were produced.[20] Moreover, Y-branched anodic alumina oxide channels were first fabricated by reducing the anodization voltage by a factor of $1/\sqrt{2}$.[21]

Similarly, after primary stem pores were fabricated by a typical two-step anodization process, n-branched nanopores were created by reducing the anodization voltage by a factor of $1/\sqrt{n}$.[22-24] Moreover, tree-like nanopore arrays were obtained by further reducing the anodization voltage by a factor of $1/\sqrt{m}$ to generate the third layer of the branced pores at the bottom of the second layer of branched pores. All the interpore distances of the above discussed porous anodic alumina were determined by the applied voltage with a linear proportional constant of 2.5 nm/V.[15] More work is needed to make porous anodic alumina with different interpore distances under the same anodization condition.

In this study, focused ion beam (FIB) patterned concave arrays are used to guide the growth of alternating-sized diameter and different nanopore arrangement. The effect of FIB guidance on enlarging the anodization window is examined. Anodization of Moiré patterns with different interpore distance is also studied in order to examine the potential of FIB patterning in fabrication of different interpore distance patterns.

EXPERIMENTAL PROCEDURE

High purity aluminum foils (99.999%, Goodfellow Corporation, Oakdale, PA) with 8 mm×22 mm×0.3mm size were used as the starting material. After being washed with ethanol and acetone, they were annealed at 500°C for 2 hrs in high purity flowing Ar gas with 5°C/min heating and cooling rates to recrystallize the aluminum foils and remove mechanical stress.

For electropolishing, the annealed aluminum foils were degreased in ethanol and acetone for 5 min, respectively, followed by DI water rinsing after each step. The aluminum foils were then immersed in a 0.5 wt% NaOH solution for 10 min with ultrasound in order to remove the oxidized surface layer. After that, the aluminum foils were electropolished in a 1:4 mixture of perchloric acid (60%-62%): ethanol (95%) (volume ratio) under a constant voltage of 12 V at room temperature with 500 rps stirring speed for 5 min.

A dual beam focused Ga$^+$ ion beam microscope (FIB, FEI Helios 600 Nano Lab, Hillsboro, OR) was employed to create different concave patterns to guide the anodization. The accelerating voltage for the FIB microscope was 30 keV. The beam diameter was ~30 nm. The beam current was 28 pA. The beam dwell time at each scan was 3 μs. The FIB created patterns were observed in the SEM mode, which allowed for in-situ monitoring of the surface features of the Al foils at different stages of the ion exposure.

The FIB patterned Al foils were anodized in 0.3 M oxalic acid at 40-60 V and 0°C for 30 min, and in 0.05 M oxalic acid at 80 V and 0°C for 30 min. For the Moiré pattern, the FIB patterned Al foils were anodized in 0.3 M phosphoric acid under 20 mA constant current at 0°C for 5 min. The voltage was ~140 V after a few seconds of anodization. Pore opening was carried out in 5 wt% phosphoric acid at 30°C for 10 min. The porous anodic alumina patterns were characterized by scanning electron microscopy (Quanta 600 FEG, FEI Company, Hillsboro, OR).

RESULTS AND DISCUSSION

FIB exposure time effect

Figure 1 shows anodized alternating diameter nanopores with different FIB patterning exposure time. The interpore distance is 125 nm. After the anodization, the pore sizes increase with the FIB exposure time even through the anodization condition is the same. When the FIB exposure time is only 1 s with the dwell time at 1 μs, both large and small FIB concaves are very shallow, less than 3 nm. The diameters of the small and large concaves are 30 nm and 65 nm, respectively. After the anodization (Figure 1(a)), the small concaves induce small anodized nanopores with 30 nm diameter. The large concaves induce large nanopores with 40 nm diameter, but the pore shape is not well defined. The nanopore arrangement maintains the pre-defined hexagonal pattern. When the FIB exposure time increases, the depth and the diameters of the concave patterns increase. For example, after 15 s of the FIB patterning, the small pores have 8 nm depth and 45 nm diameter, and the large pores have 50 nm depth and 80 nm diameter. As a result, the diameters of the anodized nanopores increase and the shapes of the anodized nanopores become much more round. When the FIB patterning time is 6 s and 15 s (Figures 1(b) and 1(c)), the anodized large nanopore sizes are 70 nm and 90 nm, respectively, and the anodized small pore sizes are 35 nm and 45 nm, respectively. At the same time, the anodized nanopores with different FIB exposure time maintain the ordered hexagonal arrangement. When the FIB time is 80 s (Figure 1(d)), the large and small anodized nanopore diameters are 90 nm and 65 nm, respectively.

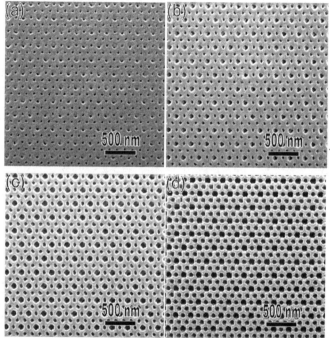

Figure 1. SEM images of anodized nanopore pattern with different FIB exposure time: (a) 1 s, (b) 6 s, (c) 15 s, (d) 80 s.

The effect of the FIB exposure time is directly related to the concave size and shape. When the FIB patterning time is short, the concaves are very shallow (Figure 1(a)), and the curvature of the concave is very large, which results in a small electrical field at the bottoms of the concaves during the anodization. The concaves cannot effectively guide the subsequent anodization; the anodized nanopores grow slowly and the pore shape is less defined. However, the Ga^+ implanation and aluminum amorphization at the concave circumferences are effective enough in maintaining the pore pattern in the hexagonal arrangement. As the FIB exposure time increases, the patterned concaves grow larger and deeper (Figures 1(b) and 1(c)), along with more Ga^+ implanation and aluminum amorphization. As a result, the anodized pores have larger diameters and more round pore shapes.

For different FIB exposure time, the pristine Al surfaces among the concaves are also different. When the FIB exposure time is short (Figures 1(a), 1(b), 1(c)), there is an un-anodized triangular aluminum pillar at the junction of a large concave and two adjoining small concaves. During the anodization, these pillars act as the effective electrical circuit to anodize nearby aluminum. Because of the asymmetric nature of the electric field, the pores are not round. The small pores elongate in the direction of the junction. As the FIB exposure time increases further (Figure 1(d)), the pristine surface after the FIB exposure diminishes/disappears, and the FIB patterning effect on the pore shape becomes less obvious or disappears. As a result, both the large and small anodized pores in Figure 1(d) are round.

Anodization voltage effect

FIB patterning also enlarges the anodization window for ordered nanopore pattern and different nanopore arrangement. Two different arrangements are studied here. The first is alternating-sized nanopores in hexagonal arrangement, and the second is uniform-sized nanopores in square arrangement. The same FIB patterned concave arrays with 150 nm interpore distance are anodized at three different conditions: 0.3 M oxalic acid at 40 V, 0.3 M oxalic acid at 60 V, and 0.05 M oxalic acid at 80 V. The SEM images of the nanopores after the anodization are shown in Figure 2. Pore widening was not undertaken on these samples and the images are the pristine structure after the anodization. As shown in Figures 2(a), 2(c), and 2(e), all these samples keep growing alternating diameter nanopores in hexagonal arrangement. As shown in Figures 2(b), 2(d), and 2(f), all the samples keep growing uniform diameter nanopores in square arrangement. However, under the traditional self-organized anodization condition, both ordered square nanopore arrangement and alternating diameter nanopores are difficult to obtain. Because the interpore distance is linearly proportional to the applied voltage with a constant of 2.5 nm/V, even with the guidance of the FIB patterning, the ordered alternating diameter and square nanopore arrangement with 150 nm interpore distance can only be synthesized at 60 V applied potential. In this study, with the FIB patterning, ordered nanopore arrangement can be anodized in the potential range of 40-80 V, which means deep concaves, Ga^+ implanation, and the re-deposition of amorphous aluminum in combination play a significant role in the nanopore growth. The FIB guidance enlarges the anodization window for obtaining the ordered nanopore arrangement.

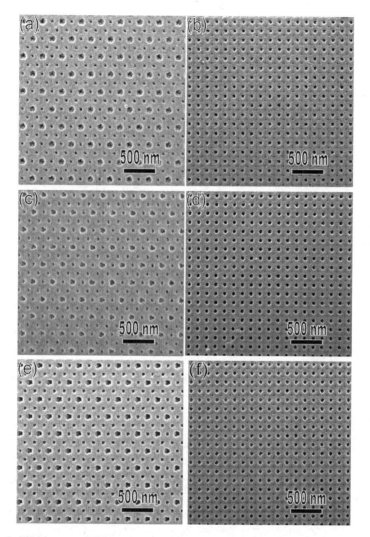

Figure 2. SEM images of anodized nanopore patterns with the same FIB pattern but different anodization condition: (a) and (b) 0.3 M oxalic acid under 40 V and 0°C for 30 min, (c) and (d) 0.3 M oxalic acid under 60 V and 0°C for 30 min, (e) and (f) 0.05 M oxalic acid under 80 V and 0°C for 30 min. The FIB pattern of (a), (c), and (e) is alternating-sized concave pattern with 150 nm interpore distance. The FIB pattern of (b), (d), and (f) is uniform concaves in square arrangement with 150 nm interpore distance.

Moiré pattern

Moiré patterns are the composite patterns created by the superposition of two identical patterns with a rotation angle, or by superposition of two different patterns with a rotation angle. There are many different interpore distances in the Moiré pattern. Since the interpore distance is dependent on the applied voltage with a linear proportional constant of 2.5 nm/V, the Moiré pattern cannot be synthesized by the conventional self-organized anodization. From the previous discussion, FIB patterning shows great potential in enlarging the anodization window and maintaining the ordered nanopore arrangement. In this section, Moiré nanopore patterns with different interpore distances are studied at a certain voltage potential.

Figure 3 shows the SEM images of anodized porous alumina arrays with graphite lattice structure Moiré patterns. The FIB patterned concave arrays are created by the superposition of two graphite lattice structure concave patterns with identical interpore distance and rotation angle of α. The interpore distance of the graphite lattice structure concave arrays is 350 nm, and the rotation angle α is (a) 5°, (b) 10°, (c) 20°, (d) 30°, respectively. After the anodization, porous alumina arrays with various Moiré patterns are obtained; the periodicity of the Moiré patterns is: (a) 6.74 μm, (b) 3.67 μm, (c) 1.60 μm, (d) 1.14 μm, respectively.

According to the fundamental theory of Moiré patterns[25-28], the periodicity D of the Moiré patterns depends on the lattice constant of both layers (d_1 and d_2) and the rotation angle α. The spectral approach, which is based on the Fourier theory, is used to analyze the Moiré phenomena. For the Moiré pattern created by the superposition of two hexagonal concave patterns with different lattice and with the rotation angle of α, the periodicity of the Moiré pattern is:

$$D = \frac{d_1 d_2}{\sqrt{d_1{}^2 + d_2{}^2 - 2 d_1 d_2 \cos\alpha}} \qquad (1)$$

When $d_1 = d_2$, the equation can be further simplified into:

$$D = \frac{d}{2\sin(\alpha/2)} \qquad (2)$$

The graphite lattice structure pattern can be considered as hexagonal pattern, as shown in Figure 4(b), and the interpore distance is $\sqrt{3}$ times of the original value. According to the spectral approach, the multiplication intensity of the resulting image is: $I(x,y) = I_1(x,y) \cdot I_2(x,y)$ in the direct space. The Moiré pattern periodicity is the distance of the two neighboring spectrum maximum. In the new hexagonal pattern, only half of the triangular gravity centers exist, such as a, b, and c in Figure 4(b). Therefore, in the direct space of Moiré pattern, the intensities of the spectrum at a, b, and c are just $I(x,y) = I_1(x,y)$, and it will not be the maximum of the resulting image. As a result, for the periodicity of the graphite lattice structure Moiré pattern in Figure 3, $d_1' = d_2' = \sqrt{3}d$, and

$D = \frac{\sqrt{3}d}{2\sin(\alpha/2)}$. From this equation, the periodicities of the Moiré patterns of Figures 3(a)-(d) are calculated as 6.95 μm, 3.48 μm, 1.75 μm. and 1.17 μm, respectively. The experimentally observed periodicities are in good agreement with the theoretical value.

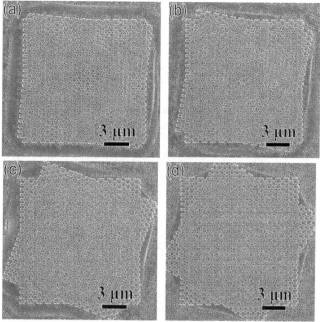

Figure 3. SEM images of porous alumina arrays with graphite lattice structure Moiré pattern. The patterns are formed by anodization of aluminum with superimposition of two identical FIB patterned graphite lattice structure concaves, the interpore distance is 350 nm, and the rotation angle between the FIB patterns is: (a) 5°, (b) 10°, (c) 20°, (d) 30°.

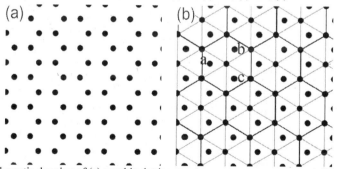

Figure 4. Schematic drawing of (a) graphite lattice structure nanopore arrangement, (b) different

view of the same graphite lattice structure arrangement as hexagonal arrangement with three points in the hexagon.

CONCLUSIONS

With the guidance of the FIB patterned concave arrays, alternating diameter nanopore arrangement is obtained. When the depth of the FIB patterned concaves is shallow, the diameter difference of the alternating diameter nanopores is not obvious. The deeper the FIB patterned concaves, the better developed the alternating diameter nanopore arrays are. The anodization window of ordered alternating diameter nanopore arrangement and square nanopore arrangement with 150 interpore distance is enlarged to 40-80 V thanks to the guidance of the FIB patterned concave arrays. This guiding effect also enables the creation of patterns with different interpore distances under the same anodization condition. Moiré pattern with various interpore distance and periodicity are obtained.

ACKNOWLEDGEMENT

The authors acknowledge the financial support from National Science Foundation under grant No. CMMI-0824741 and the Institute for Critical Technology and Applied Science of Virginia Tech. Assistance from John McIntosh and Stephen McCartney from the Nanoscale Characterization and Fabrication Laboratory of Virginia Tech is greatly acknowledged.

REFERENCES
[1] Y. Q. Wang, Y. M. Sun, and K. Li, Dye-Sensitized Solar Cells Based on Oriented ZnO Nanowire-Covered TiO$_2$ Nanoparticle Composite Film Electrodes, *Mater. Lett.*, **63**, 1102-1104 (2009).
[2] T. S. Kang, A. P. Smith, B. E. Taylor, and M. F. Durstock, Fabrication of Highly-Ordered TiO$_2$ Nanotube Arrays and Their Use in Dye-Sensitized Solar Cells, *Nano Lett.*, **9**, 601-606 (2009).
[3] D. I. Suh, S. Y. Lee, T. H. Kim, J. M. Chun, E. K. Suh, O. B. Yang, and S. K. Lee, The Fabrication and Characterization of Dye-Sensitized Solar Cells with a Branched Structure of ZnO Nanowires, *Chem. Phys. Lett.*, **442**, 348-353 (2007).
[4] J. Hahm, and C. M. Lieber, Direct Ultrasensitive Electrical Detection of DNA and DNA Sequence Variations Using Nanowire Nanosensors, *Nano Lett.*, **4**, 51-54 (2004).
[5] A. K. Wanekaya, W. Chen, N. V. Myung, and A. Mulchandani, Nanowire-Based Electrochemical Biosensors, *Electroanal.*, **18**, 533-550 (2006).
[6] Y. Cui, Q. Q. Wei, H. K. Park, and C. M. Lieber, Nanowire Nanosensors for Highly Sensitive and Selective Detection of Biological and Chemical Species, *Science*, **293**, 1289-1292 (2001).
[7] C. K. Chan, H. L. Peng, G. Liu, K. McIlwrath, X. F. Zhang, R. A. Huggins, and Y. Cui, High-Performance Lithium Battery Anodes Using Silicon Nanowires, *Nat. Nanotech.*, **3**, 31-35 (2008).
[8] K. Q. Peng, J. S. Jie, W. J. Zhang, and S. T. Lee, Silicon Nanowires for Rechargeable Lithium-Ion Battery Anodes, *Appl. Phys. Lett.*, **93**, 033105-1 (2008).
[9] H. Asoh, K. Nishio, M. Nakao, A. Yokoo, T. Tamamura, and H. Masuda, Fabrication of Ideally Ordered Anodic Porous Alumina with 63 nm Hole Periodicity Using Sulfuric Acid, *J. Vac. Sci. Technol. B*, **19**, 569-572 (2001).
[10] H. Masuda, F. Hasegwa, and S. Ono, Self-Ordering of Cell Arrangement of Anodic Porous

Alumina Formed in Sulfuric Acid Solution, *J. Electrochem. Soc.*, **144**, L127-L130 (1997).

[11]A. P. Li, F. Muller, A. Birner, K. Nielsch, and U. Gosele, Hexagonal Pore Arrays with a 50-420 nm Interpore Distance Formed by Self-Organization in Anodic Alumina. *J. Appl. Phys.*, **84**, 6023-6026 (1998).

[12]F. Y. Li, L. Zhang, and R. M. Metzger, On the Growth of Highly Ordered Pores in Anodized Aluminum Oxide, *Chem. Mater.*, **10**, 2470-2480 (1998).

[13]H. Masuda and K. Fukuda, Ordered Metal Nanohole Arrays Made by a Two-Step Replication of Honeycomb Structures of Anodic Alumina, *Science*, **268**, 1466-1468 (1995).

[14]H. Masuda, K. Yada, and A. Osaka, Self-Ordering of Cell Configuration of Anodic Porous Alumina with Large-Size Pores in Phosphoric Acid Solution. *Jn. J. Appl. Phys. Part 2-Letters*, **37**, L1340-L1342 (1998).

[15]K. Nielsch, J. Choi, K. Schwirn, R. B. Wehrspohn, and U. Gosele, Self-Ordering Regimes of Porous Alumina: The 10% Porosity Rule, *Nano Lett.*, **2**, 677-680 (2002).

[16]H. Asoh, S. Ono, T. Hirose, M. Nakao, and H. Masuda, Growth of Anodic Porous Alumina with Square Cells, *Electrochim. Acta*, **48**, 3171-3174 (2003).

[17]N. Y. Kwon, K. H. Kim, J. Heo, and I. S. Chung, Fabrication of Ordered Anodic Aluminum Oxide with Matrix Arrays of Pores Using Nanoimprint, *J. Vac. Sci. Technol. A*, 27, 803-807 (2009).

[18]Z. P. Tian, K. Lu, and B. Chen, Unique Nanopore Pattern Formation by Focused Ion Beam Guided Anodization, *Nanotechnology*, accepted.

[19]H. Masuda, H. Asoh, M. Watanabe, K. Nishio, M. Nakao, and T. Tamamura, Square and Triangular Nanohole Array Architectures in Anodic Alumina, *Adv. Mater.* **13**, 189-192 (2001).

[20]B. Chen, K. Lu, and Z. P. Tian, Gradient and Alternating Diameter Nanopore Templates by Focused Ion Beam Guided Anodization, *Electrochim. Acta*, accepted.

[21]J. Li, C. Papadopoulos, and J. Xu, Growing Y-Junction Carbon Nanotubes, *Nature*, **402**, 253–254 (1999).

[22]G. W. Meng, Y. J. Jung, A. Y. Cao, R. Vajtai, and P. M. Ajayan, Controlled Fabrication of Hierarchically Branched Nanopores, Nanotubes, and Nanowires, *Proc. Nat. Acad. Sci.* **102**, 7074 (2005).

[23]S. S. Chen, Z. Y. Ling, X. Hu, and Y. Li, Controlled Growth of Branched Channels by a Factor of $1/\sqrt{N}$ Anodizing Voltage?, *J. Mater. Chem.*, **19**, 5717–5719 (2009).

[24]J. P. Zhang, C. S. Dayb, and D. L. Carroll, Controlled Growth of Novel Hyper-Branched Nanostructures in Nanoporous Alumina Membrane, *Chem. Commun.*, **45**, 6937–6939 (2009).

[25]J. Choi, R. B. Wehrspohn, and U. Gösele, Moiré Pattern Formation on Porous Alumina Arrays Using Nanoimprint Lithography, *Adv. Mater.*, **15**, 1531-1534 (2003).

[26]V. Luchnikov, A. Kondyurin, P. Formanek, H. Lichte, and M. Stamm, Moiré Patterns in Superimposed Nanoporous Thin Films Derived from Block-Copolymer Assemblies, *Nano Lett.*, **7**, 3628-3632 (2007).

[27]I. Amidror, Moiré Patterns Between Aperiodic Layers: Quantitative Analysis and Synthesis, *J. Opt. Soc. Am. A*, **20**, 1900-1919 (2003).

[28]I. Amidror, The Theory of the Moiré Phenomenon, Kluwer *Academic Publisher:Norwell*, MA, (2000).

THIN FILMS OF TiO$_2$ WITH Au NANOPARTICLES FOR PHOTOCATALYTIC DEGRADATION OF METHYLENE BLUE

F. Palomar[1], I. Gómez[1]*, J. Cavazos[2]

[1]Laboratorio de Materiales I, Facultad de Ciencias Químicas, UANL, San Nicolás de los Garza, N.L., México

[2]Departamento de Mecánica, Facultad de Ingeniería Mecánica y Eléctrica, UANL, San Nicolás de los Garza, N.L., México

*mgomez@fcq.uanl.mx

ABSTRACT

Au/TiO$_2$ thin films have been prepared from Titanium Tetrabuthoxide-acetylacetone solution in buthanol as solvent by sol-gel and dip-coating technique. Au nanoparticles were prepared from HAuCl$_4$ solution. The deposition of Au nanoparticles was by spray pyrolysis method. The films from the sol gel solution were heat treated at 450°C by 1 hour. The surface structures, morphology, composition and optical properties on the films were investigated by atomic force microscopy (AFM), optical microscopy (OM), X-ray Diffraction (XRD) and UV-Vis spectrometer. It was found that the film consisted of anatase phase (TiO$_2$), Au nanoparticles in the range of 20 to 80 nm were deposited. In this study, we measured the photocatalytic degradation of a methylene blue (MB) aqueous solution by the thin films prepared. We found that Au nanoparticles deposited on the TiO$_2$ films improved the photocatalytic activity of the TiO$_2$ films and diminished degradation time.

INTRODUCTION

Titanium dioxide (TiO$_2$) is a well-know photocatalyst material. When titanium oxides are irradiated with UV light with energy that is greater than the band gap energy of the catalyst (about λ=380 nm), electrons (e-) and holes (h+) are produced in the conduction and valence bands, respectively. These electrons and holes have a high reductive potential and oxidative potential, respectively, which, together, cause catalytic reactions on the surfaces; namely photocatalytic reactions are induced. In the presence of O^2 and H$_2$O, the photo-formed e- and h+ react with these molecules on the titanium oxide surfaces to produce O^{2-} and OH radicals, respectively. These O^{2-} and OH radicals have a very high oxidation potential, inducing the complete oxidation reaction of various organic compounds[1]. TiO$_2$ under anatase form is usually used as an efficient photocatalyst.

Several attempts, for example, ion doping[2], noble metal deposition[3] and adding a coadsorbent[4] are mostly used to enhance this activity. Ion doping and noble metal deposition can separate photogenerated electrons and the holes; while the function of coadsorbent is to improve the absorption of the catalyst. The approach of establishing a semiconductor-metal composite system is usually considered as an effective way to reduce the recombination of electrons and holes and improve quantum efficiency of the photocatalytic process[5]. Metal particles such as Au and Ag have been reported to enhance the photocatalytic performance of TiO$_2$, because the photogenerated electrons are trapped by these particles leading to high efficiency of charge separation[6].

Spray pyrolysis involves passing a solution or suspension of precursors through an aerosol generator to form droplets suspended in a gas. The droplets are then heated by passing the aerosol through a tube contained in a furnace, where thermally induced chemical reactions in the particles or reaction with gaseous co reactants convert the precursors to the product powder, which is either collected or deposited on a substrate[7]. The aerosol can be generated by liquid atomization (with high velocity air) or ultrasonic atomization (without air), the second has the advantage of narrow drop size distribution and hence, narrow particle size distribution[8].

In this work, ultrasonic spray pyrolysis was used for the deposition of gold nanoparticles on a glass substrate and using TiO$_2$ as a surrounding medium to increase the substrate affinity of the gold nanoparticles. Three different flow rates of the carrier gas were used. The optical properties and the morphology were studied and the particle size was measured by scanning electronic microscopy and atomic force microscopy. The objective of this work was to study the effect of adding Au nanoparticles by spray pyrolysis deposition on thin films of TiO$_2$ and comparing degradation of methylene blue dye.

EXPERIMENTAL PROCEDURE

Materials
Glass CORNING with thickness of 0.8-1.1mm, Titanium Tetrabuthoxide analytic grade 97% Sigma-Aldrich, Buthanol J.T. Baker 99.96%, 2,4-pentanodione 99% Aldrich, were used to prepare the thin films.

Solution Preparation
A solution of 42mL of titanium tetrabuthoxide was diluted in 13 mL buthanol, after 25 mL of 2,4-pentanodione was added under constant stirring.

Thin Films
Deposition of thin films was carried out by Dip-Coating method[8], the substrate, which was previously washed with water and dried with air, was submerged in the titanium solution with adjusted pH with acetic acid, at a rate of 1cm/min, after that, heat treatment at 450°C for 1 hour was applied. These procedures were repeated for 1, 5, 10, 15 and 20 layers.

Au Nanoparticles Preparation
0.5g of gold powder were mixed with 4 mL of a mixture of hydrocloric and nitric acid, then heated to boiling point for the volume to be reduced to 0.75 mL. Distilled water was added to dilute to 25 mL to obtain a 100 millimolar gold solution and then diluted to 0.1 millimolar; after that the solution was heated to boiling point and sodium citrate was immediately added.

Au Nanoparticles Deposition
The Au solution was placed in the ultrasonic nebulizer (frequency = 1.7 MHz). The flow rate of the carrier gas (N$_2$) used was 2 L/min. The glass substrate with thin films of TiO$_2$ was placed inside a quartz tube, which was contained in a tubular furnace heated at 500 °C. The aerosol was passed through the quartz tube. With the heating the HAuCl$_4$ decomposes according the equation 1, this reaction occurs at 175 °C.

$$HAuCl4 \rightarrow Au(s) + HCl\ (g) + 3/2Cl2(g) \qquad (1)$$

Photocatalysis

In order to prove the effect of photocatalysis solutions of methylene blue dye were prepared at concentrations of 1 ppm, then the coated substrates were placed inside the containers with the solution of methylene blue, and the containers were placed inside a black box with a 365 nm UV lamp to 60W, while maintaining a distance of 8 cm between the substrates and the UV lamp. Samples were taken every 30 minutes to measure the absorbance by UV-Vis spectroscopy.

Characterization

X-ray diffractometer Rigaku D. Max 2000, with Cu-Kα radiation was used to obtain the diffraction patterns. In order to calculate the forbidden energy band (Eg) from deposited TiO₂, the spectra were recorded on UV-Vis spectrophotometer Perkin Elmer Lambda 12, measurements from 200 to 700 nm were made and the reference was the absorbance of the glass substrate.

Analysis by optical microscopy was performed on an Olympus BX60 microscope. Analysis by Atomic Force Microscopy was made in Quasant Model Q-Scope 3500 in tapping mode.

DISCUSSION

X-Ray Diffraction

The samples prepared with TiO₂ were characterized by X-ray diffraction. Figure 1 presents the spectrum of the substrate covered with TiO₂. In the diffractogram are observed the reflection peaks characteristical of the anatase phase present in the sample, which correspond to those reported in the literature[7].

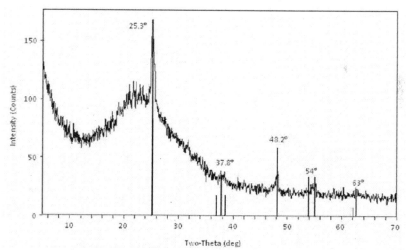

Figure 1.Difractogram of TiO₂ identifying the anatase phase.

Ultraviolet-Visible Spectroscopy
    Analysis of Ultraviolet-Visible spectroscopy determined the energy of forbidden band (Eg) for the different numbers of layers on the substrate (Figure 2). Results are shown in Table 1 where it can be seen that the values of the energy of forbidden band of samples prepared with Au nanoparticles present variation, because the Au nanoparticles produced are capable of transferring charge to the semiconductor and thus making a more efficient use of sunlight[6].

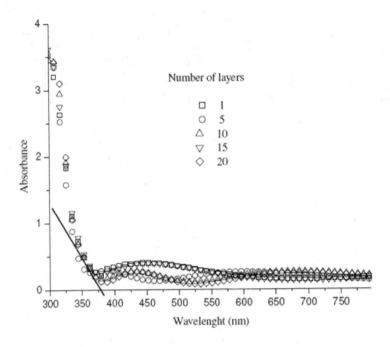

Figure 2. UV-Vis curves of thin films with different numbers layers.

Atomic Force Microscopy
    From the AFM analysis it is shown that the gold nanoparticles have been deposited on the TiO$_2$ film with sizes ranging from 50 to 100 nm. In this figure we can see the nanoparticles deposited as protuberances with elongated morphology such as spheroids, with growth according to the gas flow direction from the spray-pyrolysis system. See Figure 3.

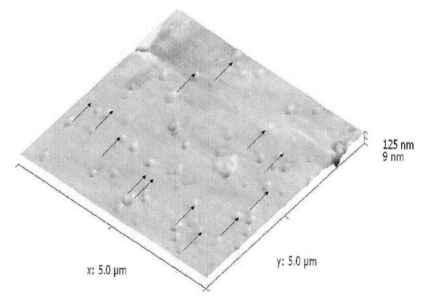

125 nm
9 nm

x: 5.0 µm

y: 5.0 µm

Figure 3. Micrograph of gold nanoparticles on the TiO$_2$ thin film. (Arrows indicate the flow direction of the carrier gas from the spray-pyrolysis system used).

Photocatalysis

Figure 4 shows the curves for degradation of the methylene blue dye solution when the solution were photocatalytically degraded by the TiO$_2$ and Au/TiO$_2$ deposited. The Au/TiO$_2$ deposited films improve the photocatalytic activity, the Au nanoparticles enhance the photocatalytic performance of TiO$_2$, because the photogenerated electrons are trapped by these nanoparticles leading to high efficiency of charge separation electron-hole.

Table I. Comparation of substrates with different number layers.

Number Of Layers	Composition	% Degradation Methylene Blue	Degradation time (min)	Forbidden Band (eV)
1	TiO$_2$	73	240	3.58
5	TiO$_2$	59	240	3.58
10	TiO$_2$	63	240	3.58
15	TiO$_2$	88	240	3.58
20	TiO$_2$	64	240	3.58
15	Au/TiO$_2$	89	120	3.12

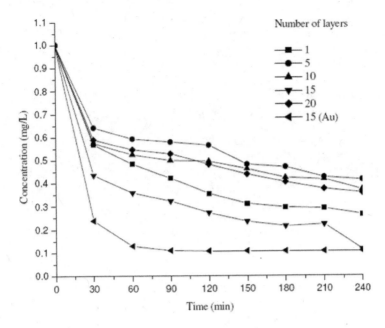

Figure 4. Curves of degradation of methylene blue with TiO$_2$ and Au/TiO$_2$.

CONCLUSIONS

According to the results of X-ray diffraction performed on samples, it was found that the TiO$_2$ deposited corresponds to the structure of anatase phase. The atomic force microscopy shows the size of nanoparticles obtained to be in the range of 50 to 100 nm. From UV-Vis spectra forbidden band energies (Eg) of TiO$_2$ and Au/TiO$_2$ films for different numbers of layers deposited were determined, from 3.58 and 3.12 eV, respectively. We found that the values for the energy of forbidden band of samples prepared with Au nanoparticles present variation in the energy of forbidden band. As this value decreases, we could use visible light or even sunlight.

From the results of photocatalysis it can be seen that the methylene blue is degraded by 89% with TiO$_2$ and by 88% with Au/TiO$_2$, which is attributed to the photogenerated electrons trapped by these nanoparticles leading to a high efficiency of charge separation electron-hole. All of this is done in half the time required by TiO$_2$ without Au nanoparticles.

ACKNOWLEDGMENTS

Authors express their gratitude to PAICYT and CONACYT for the financial support to this work and the Materials Laboratory from FCQ-UANL.

REFERENCES

[1]V. Ramamurthy and Kirk S. Schanze, *Mol. and Supramol. Photochem.*, **8**, (2001).
[2]Wood. A. Giersing and M. Mulvaney, *J. Phys. Chem. B*, **6**, 837-54 (2008).

[3]M. Sadeghi, W. Liu, T. Zhang, P.Stavropoulos and B. Levy, *J. Phys. Chem.,* **50**, 19466-74 (1996).
[4]C. Cai, J. Zhang, F. Pan, W. Zhang, H. Zho and T. Wang, *Catal. Lett.,* **123**, 51-55 (2008).
[5]N. Shahina and H. M. Fabrés, Ahmed, *Bull. Matter. Sci.,* **31**, 43-48 (2008).
[6]C. Yogi, K. Kojima, T. Takai and N. Wada, *J. Mater. Sci.,* **44**, 821-27 (2009).
[7]W. Xianyu, M. Park and W. Lee, *Kor. J. Chem. Eng.,* **18**, 903-07 (2001).
[8]M. Abdul and I. Rahman, *Mal. J. Chem.,* **5**, 86-91 (2003).

NEW ENTROPIC ROUTES FOR NANO-BANDS AND NANO-PARTICLES

H.P. Li (hli@just.edu.tw)
Jinwen University of Science and Technology,
Hsintien, New Taipei City, 23154, Taiwan

G.K. Dey (gautam_k_dey@yahoo.com)
Bhabha Atomic Research Centre,
Trombay, Mumbai 400 085, India

J.A. Sekhar (sekharja@ucmail.uc.edu)
Department of Chemical and Materials Engineering, University of Cincinnati,
College of Engineering, Cincinnati, OH 45221- 0012, U.S.A

ABSTRACT

Following our recent publications, a relationship is inferred between dissipative reactions, nano-crystal formation, and nano-bands in micropyretically synthesized equimolar NiAl alloys. Time-lapse X-ray reports, microstructural studies, process conditions, and combustion calculations are correlated for understanding the micro-kinetics of synthesis. Various micro-kinetic mechanisms may be operative depending on the chosen initial conditions and alloy chemistry. Dissipative oscillatory chemical reactions, called Belousov-Zhabotinsky (BZ) reactions, are proposed as one synthesis mechanism, which leads to the formation of the observed nano-scale features such as nano-particles and nano-bands. Nano-bands in solid state combustion processes are discussed. The dissipative oscillations, that are a consequence of the non-linear reaction rate equations, create and simultaneously disperse nano-particles and nano-bands depending on the initial temperature, composition, and the process conditions chosen. The spatio-temporal structure from a moving geometrical configuration such as a micropyretic solid-state combustion front can contain a decaying dissipative reaction product, *e.g.*, a combustion spin microstructure. Nano-band forming waves and nano-crystals possibly interact leading to unique variations in the microstructure. Such microstructural possibilities could be advantageously controlled by manipulating the initial conditions. The implications of the BZ finding could be significant; as it offers a method of forming bulk near-net shaped objects containing nano-structured enhancements.

INTRODUCTION

We show in this article that the BZ reaction can leave a residue of nano-particles and very fine faulted bands (nano-bands), which are distributed in the cooled alloy grains [1-9]. Micron-scale bands and precipitates are already recognized in low temperature BZ chemical reactions [7,8]. The distribution and length scale of the precipitates correspond to the wave structure of the propagating chemical waves, *e.g.*, micron size structures in room temperature silica gel and silver nitrate reaction mixture [7]. Inside the room temperature wave front (width 180-200 microns) an additional thin precipitation band (width 10-40 micron) of dense micron size precipitates was also observed [7]. In this article, we identify equivalent particles and bands, however, with a nano-scale structure for the higher temperature, more rapid micropyretic reactions. The traveling wave in a high temperature combustion reaction offers oscillatory BZ dissipation possibilities as has been recently noted [1]. Dissipation has diffusive origins with a cyclic hysteresis leading to directional diffusion conditions (*e.g.*, the spin condition identified in reference 1). The main evidence of the BZ reaction in reference 1, came from the analysis of the micron-scale repetitive thin banded structures in high aluminum content Ni-Al alloys (*e.g.*, between the Ni_2Al_3 and $NiAl_3$ compositions), that were micropyretically synthesized by an one-dimensional combustion wave propagation condition. Although there is no dearth of amazing properties reported on nano-structured alloys [10-15], there are *three* main difficulties for making reliable, uniform, nanostructure bulk alloys. These are: **(a) Microstructural non-uniformity:** This is an issue that manifests particularly when starting with nano-powders for the bulk object manufacture by powder metallurgy routes. Uniform microstructures are a near impossibility because of the agglomeration and non-uniform coarsening during sintering of agglomerated nano-particles. The problems encountered with employing a conventional powder metallurgy synthesis route for nano-particle consolidation are also compounded by irregular nano-particle sizes and shapes which then lead to non-uniform packing morphologies that result in packing density variations in the powder compact. Such uncontrolled agglomeration of powders often arises from attractive van-der-Waals forces between the particles and therefore is worse as the chosen particle size(s) decrease. A review by Zhang et al. [16] illustrates the problem of rapid agglomeration and low temperature sintering of nano-materials including the differential stresses that arise from inhomogeneous densification that cause internal cracks. **(b) Energy wastage that impacts the cost of production:** Nano-structured alloys are also made sometimes with continuous high energy impact deformation processing. Examples of such processes are rapid forging and multiple pass-extrusions which all use *extremely* high amounts of energy [9,17]. Apart from the serious energy penalty that such processes impose by this brute force approach, there is also the cost of dies, the high cost of heating (to soften the alloys) and a myriad of associated capital equipment costs and safety issues. **(c) Human safety with nano-particles:** Nano-particles are often thought to negatively impact human health and the environment [18-22]. While nano-materials and nano-technologies are expected to yield numerous healthcare advances such as more targeted methods of delivering drugs, new cancer therapies, and methods of early detection of diseases, they also may have unwanted effects. The high specific surface area (surface area per unit weight) of nano-particles compared to micron size particles may lead to an increased rate of absorption through the skin, lungs, eyes, or digestive tract. As the use of nano-materials increases, the worldwide concerns for worker and user safety will increase even more. There are numerous ongoing programs that are directed at avoiding nano-particles during synthesis and handling [23]. Micropyretic synthesis has been shown to yield nanostructures but there are no reports which identify the nanostructure yielding conditions.

Nano scale grains are now clearly identified even in the single phase regions in NiAl and Ni_2Al_3 compositions [24-30]. Dey and Sekhar [31] have shown nano-particles residing (without any orientation relationship to either grain) at low angle grain boundaries in NiAl alloys with small amounts of Fe, V, Ti additions (*i.e.*, in dilute NiAl alloys). More recently Curfs *et al.* [24,26,27] and Tolochko *et al.* [32] have identified nano-scale features in unalloyed NiAl, *i.e.*, without any additional solute additions. Nano-scale microstructural features have also been noted in micropyretically synthesized FeAlMn alloys [33]. With an increase in the Mn content, the size of the nano scale features decreased up to 10 wt.% Mn addition. The room temperature fracture toughness appears to be impacted by the nano-bands and nano-particles. In the FeAlMn alloys [33], the fracture toughness increased with the diminishing size of the nano scale faults.

The possibility for obtaining alloys containing nano-features, through net-shape *bulk-processing* routes (such as micropyretic routes [34] or controlled thermomechanical deformation), has advantages that have not been explored previously for most materials, exceptsteels [35-39]. Soft bands and nano-particles in steels have been proposed for making improved ductile steels, but processing limitations have prevented wide scale use [35-39]. Controlled micropyretic synthesis offers a similar property control for NiAl, and possibly for other compounds [34]. However perplexing questions regarding microstructural uniformity plague the use of micropyretically synthesized alloys. The problem is illustrated by an example from NiAl alloys. A summary of reported mechanical properties [31,40,41] and phase constituents of micropyretically synthesized NiAl alloy objects is shown in the previous study [9] and Table 1 [10,27,29,42-47]. The increase in compressive ductility and increase in toughness are not fully correlated. In fact, the published data demonstrates a significant lack of correlation of mechanical properties with variations in the NiAl alloy chemistry and microstructure leading to an inference that micropyretically synthesized alloys could show erratic properties. In this article, we also consider the basis for the differences in toughness as originating from hitherto unknown BZ oscillations and propose a processing method to control these.

It is important to distinguish between the micro-kinetics and macro-kinetics of micropyretic synthesis. The microstructural features can be independently influenced by the macro-kinetics (*i.e.*, the conditions that impact overall heat flow and combustion wave velocity) as well as by the micro-kinetics (*i.e.*, BZ reaction and initial particle-scale local transport conditions discussed below). For example, macro-kinetics and thermal gradients determine segregated porosity from Lewis type oscillations [1]. The micro-kinetics are influenced by alloy chemistry and other imposed *initial* conditions, and under certain conditions give rise to dissipative waves [1]. The scale of microstructure that is influenced by either type of kinetics could be on vastly different but possibly also interact in unanticipated ways. Such interactions could explain properties like toughness that are presently uncorrelated to porosity or chemistry in micropyretic synthesis. Although porosity can sometimes increase toughness [48,49], the large amount of angular porosity noted in the micropyretically synthesized alloys, is not expected to enhance the toughness. Yet toughness increases are sometimes noted. Except for the possibility of higher ductility with solute additions that alter the lattice parameter [50-53], no satisfactory explanation exists for the reported large increase in toughness with the small solute additions or by initiating combustion at intermediate initial temperatures. For example, Fe which shows little preference for Ni or Al sites [54] and reduces the lattice parameter, significantly increases the ductility, but only somewhat increases the toughness. Solutes such as Cr, Ti, V, and Nb do not increase ductility to the same degree but significantly enhance toughness under certain conditions [31,40,54]. The fracture toughness of NiAl objects, containing very small amounts of ~0.25 at. % Fe or containing other small amounts of solutes such as Cr, Ti, V, and Nb [31,40,41,55], has been reported to be 12.5 MPa.m$^{1/2}$ when synthesized micropyretically [40]. The value is very high, especially considering the amount of angular porosity (~20 vol. %) in the objects, and much higher than the best values reported in by others [50,51,56-59] for alloyed high-density NiAl. Without any alloying, the reported values are about 6 MPa.m$^{1/2}$, for a high-density NiAl alloy and also for the same composition and high purity *porous* micropyretically synthesized alloys.

The high frequency BZ oscillation that are required for nano scale features can only proceed when a bifurcation parameter is exceed if the sub-reaction sequence is able to support oscillating compositions [1]. A Brusselator sequence (a simple BZ sequence) is discussed below for BZ oscillations in the equimolar NiAl system. The sub-reactions shown below, especially when oscillatory (*i.e.*, dissipative), offers a method of dispersing nano-particles and faulted regions in a manner similar to micro-particle and micro-band formation in lower temperature chemical reactions [7]. We also compile evidence from the literature for the sub-reactions and BZ hypothesis from published synchrotron and fine-scale time-lapse XRD studies [24,26,27,32,46,60,61] for explosive micropyretic synthesis of NiAl and other compounds. A compilation of published data [24,25,41,47,62-65] and new data on thermal oscillations also points to the BZ hypothesis [31,40,66-71].

Table 1 Variation in the product phases generated from the different processing techniques and initial conditions [10,27,29,42-47]. Note the change in the observed phases with a change in the initial temperature.

Ni:Al	Process Method *	Initial Temperature	Product Phases	Ref.
1:1	SHS	25°C	NiAl	47
1:1	SHS	25°C	NiAl	46
1:1	MA	25°C	NiAl	10
1:1	HERS	300°C	NiAl	44
1:1	HERS	375°C	NiAl	44
1:1	TE	660°C	NiAl Ni$_2$Al$_3$, NiAl$_3$, Ni$_5$Al$_3$,	27
1:1	TE	660°C	NiAl + other phases	42
1:1	HPRS	900°C	NiAl, Ni$_2$Al$_3$, Ni$_3$Al,	45
1:1	MA	970°C	NiAl NiAl$_3$, Ni$_3$Al$_4$,	29
1:1**	RHP	1250°C	NiAl + other phases	43

* HERS (hot extrusion reaction synthesis); HPRS (high pressure reaction sintering); RHP (reaction hot pressing); MA (mechanical alloying), TE (thermal explosion mode), SHS (self-propagating high-temperature synthesis).
** with 20 vol.% NiAl diluents.

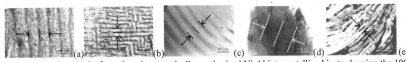

Figure 1 (a) A micrographs from the micropyretically synthesized Ni-Al intermetallic objects showing the 100 μ m band feature [72]. (b) Bright-field image of the wedge-shape specimen in a (β'-NiAl alloy) annealed at 623K for 1 hour showing the 0.002 μ m martensitic type band feature [74]. (c) A TEM micrograph showing the 0.1 μ m features noted in the as-synthesized high purity unalloyed NiAl material [40]. (d) High resolution images of mechanically alloyed (milled for 24 hrs) nano-crystalline NiAl particles displaying a 0.1 μ m band feature [40]. (e) The deformation induced or mechanically activated band structured reported in equimolar NiAl alloys with a 0.1 μ m band feature [62].

BANDED STRUCTURES IN NiAl.

At least five types of banding features have been reported in the literature on equimolar NiAl alloys. The largest of these are the 100 μ m bands shown in references 1, 32, 72, and 73 that are commonly noted in micropyretically synthesized alloys. A typical 100 μ m band feature is shown in Fig. 1(a) [from reference 72]. These are the type of bands which were discussed in detail in reference 1 and belong to a combination of Lewis and BZ bifurcations [1]. The smallest of the reported bands is the martensitic NiAl (beta-NiAl) (Fig. 1(b)) which normally has a < 0.002 μ m lath size [74].

The rest of the three identified bands lie in the 0.1 μ m - 1 μ m range and possibly have similar micro-kinetic origins for their synthesis regardless of the work-based (forging and extrusion) or heat-based synthesis (micropyretic synthesis or casting) techniques. These include (1) the curved band faults (Fig. 1(c)), which are noted in the range of 0.1 μ m - 1 μ m.. Such bands form during high· temperature micropyretic synthesis in the thermal explosion or unidirectional mode [31,40,41]; (2) the shear bands which have been identified as a band type feature that forms during mechanical deformation [10] (Fig. 1(d)); and (3) the deformation induced or mechanically activated bands shown in Fig. 1(e) [29,45,62]. The widths of shear bands or mechanically activated bands are also in therange of 0.1 μ m to 1 μ m [10,62,75]. In this article, we discuss one type of BZ kinetics which can lead to 0.1 μ m type of fault bands.

BRUSSELATOR MODEL FOR NiAl

A Brusselator model for high aluminum containing nickel alloys was developed in reference 1. This model only involved two intermediaries and was unable to describe any oscillations in equimolar NiAl although the model appeared adequate for describing the compositions in the higher aluminum range i.e. closer to Ni_2Al_3 and $NiAl_3$. A new modified Brusselator model which shows oscillations for four intermediaries A', A'', B', and Y is constructed as shown below in Eqs. (1a) – (1e) to describe a chemical oscillator for NiAl. This sequence will occur after a bifurcation condition equation (2) is exceeded [6]. The conventional single step overall irreversible entropy producing reaction with two reactants, i.e., A+B → (D+E), is still possible prior to the bifurcated sequence. Other bifurcations may also exist with a higher number of sub reaction sequences but the five step sequence is adequate to explain the NiAl microstructures reported to date [9]. The sequence takes the form:

$$A \rightarrow A' \qquad <k_{1f}> \qquad (1a)$$
$$B + A' \rightarrow B'+Y \qquad <k_{2f}'> \qquad (1b)$$
$$B'+1/2A' \rightarrow D \qquad <k_{2f}''> \qquad (1c)$$
$$2A'+ Y \rightarrow 3A'' \qquad <k_{3f}> \qquad (1d)$$
$$3A'' \rightarrow E+5/2A' \qquad <k_{4f}> \qquad (1e)$$

The k's indicate the dominant forward reaction rate constants. All forward reaction rate constants have the dimensions of 1/s when the concentrations are expressed in a dimensionless form. Similar reactions can be written for compositions upto $NiAl_3$, between $NiAl_3$ and Ni_2Al_3 and between Ni_3Al and NiAl.

Oscillations can be obtained in the concentration of the intermediary species B' and Y indicating dissipative cyclic phenomena [4], provided that the critical bifurcation condition is exceed (Eq. (2) below) where k_{2f}' and k_{2f}'' is reduced to one reaction rate k_{2f}. This condition is exceed when the dimensionless concentration of species B, expressed as [B], is greater than the RHS of the bifurcation condition in Eq. (2), i.e., when,

$$[B] > \frac{k_{4f}}{k_{2f}} + \frac{k_{3f}k_{1f}^2}{k_{2f}k_{4f}^2}[A], \qquad (2)$$

Here k_{1f} to k_{4f} are the forward reaction rate coefficients for the Eqs. (1a)-(1e) respectively when k_{2f}' and k_{2f}'' are clubbed into one reaction k_{2f}. A typical oscillation periodicity is in the order of milliseconds for a $k_{2f}>1000$ [1]. Most of the auto-sustaining micropyretic reactions happen well above 1500K [75] whenever there is a high reaction rate constant, i.e., for $k_{2f}>1000$ [1]. Micropyretic synthesis is associated with a moving combustion front whether from multiple local ignitions points in a compact, or from a two dimensional unidirectional ignition condition [34]. When considering a spatially moving front, such as a combustion front, the concentrations of A and B are maintained at non-equilibrium levels by flows of reactants which can be continuously enabled by the unreacted part of the overall body of reactants as the combustion wave propagates. Some of theomplexities arising from defect chemistries and additional sub-steps have been discussed in reference 1. In Eqs. (1a)-(1e) A' and A'' are closely related compositions such as for example from a Ni melt with variations in aluminum concentration. The intermediate A'' is a compositional variation in A' such as a solution with increasing solute content. If E is able to absorb some stoichiometric variations in order to create A' (Eq. (1e)), then regardless of the composition of A'' a classic BZ may be inferred, although the sequence will show decaying concentrations and possible loss of BZ when the bifurcation condition Eq. 2, is again changed. The concentration of intermediate A'', $i.e.$, [A''], could also change with time from [A'] and set up a slow decay oscillation (non steady-state conditions than can change the bifurcation regime). Oscillation in [B'], [Y] and decay of [A''] will be noted when plotted against time for a fixed location similar to the more recognized four sequence Brusselator [1,6].

The substitution of reactants (A, B) and intermediaries (A', A'', B' and Y) in Eq. (1) by $Al_{(s)}$ (with the small amount of solute Fe), $Ni_{(s)}$, $Al(Fe)_{(l)}$, $Al(Ni,Fe)_{(l)}$, $Ni_{(l)}$, and $[(Ni_2Al_3+NiAl_3)(Fe)_z]_{(r,s)}$ may be made. The product substitution $D=NiAl(Fe)_{(r,s)}$ and $E=NiAl(Fe)_{(s)}$ can now be made which can show oscillations of the BZ type in the following fashion.

$Al_{(s)}$(heated) $+0.004Fe \rightarrow Al(Fe)_{(l)}$ (3a)

$Ni_{(s)} +Al(Fe)_{(l)} \rightarrow 1/2Ni_{(l)}$(band initialization) $+ 1/6[(Ni_2Al_3+NiAl_3)(Fe)_z]_{(r,s)}$ (3b)

$1/2Ni_{(l)} +1/2Al(Fe)_{(l)} \rightarrow 1/2NiAl(Fe)_{(r,s)}$ (3c)

$2Al(Fe)_{(l)} + 1/6[(Ni_2Al_3+NiAl_3)(Fe)_z]_{(r,s)} \rightarrow 3Al(Ni,Fe)_{(l)}$ (3d)

$3Al(Ni,Fe)_{(l)} \rightarrow NiAl(Fe)_{(s)}+5/2Al(Fe)_{(l)}$(fault nano-bands form along with (3c))(3e)

The combinations of Equation (3b), (3c) and (3d) are the classic Brusselator. The reactions can happen sequentially or in an oscillatory manner $i.e.$, in a dissipative BZ form. Here the subscript s indicates solid, l indicates liquid, and r indicates a reaction product. When used together r,s indicates a reaction product/compound in the solid phase. The brackets indicate the presence of dissolved solute. The subscripts x and y reflect the fact that a defect structure may be included in the sub reaction sequences.

This is the first report of nano-bands in solid state combustion processes. The predicted band spacing based on Eqs. (3b)-(3d) is 1:1.5 ($D=NiAl(Fe)_{(r,s)}$: $E=NiAl(Fe)_{(s)}$). Such a ratio is noted for NiAl bands noted in reference 41. Particles that are less prone to dissolution compared to Ni_2Al_3 and $NiAl_3$ (Eq. (3d)) may not be fully dissolved and can appear at the fault (band) boundaries or grain boundaries. For example, vanadium containing particles (with a much higher melting point (see data in reference [76]) compared to Ni_2Al_3 (1133°C) and $NiAl_3$ (854°C)) may be inferred at the band boundaries if Eqs. (3d) is rewritten in the form:

$2Al(Fe)_{(l)} + 1/6[(Ni_2Al_3+NiAl_3)(Fe)_z]_{(r,s)} + 0.001V \rightarrow 3Al(Ni,Fe)_{(l)} +$

$0.005/6[V_2NiAl_x]_{(r,s)}$ (when vanadium is present as a solute) (3d')

Many self-organizing systems show at least one bifurcation. The proposed sequence above is indicative of more than one bifurcation possibility. Depending on the outcome that is studied, bifurcations are possible which impact different length scales of the microstructure considering, e.g. the Lewis instability [1] and solidification interface instabilities [77]. These larger scale features are from what we term as macro-kinetic (like the combustion velocity, temperature profile) type bifurcations. Although not well understood yet these bifurcations may interact with the micro-kinetic bifurcations.

An increase in the initial temperature (T_o) will increase the reaction temperatures and could prevent the onset of possible BZ bifurcations. The overall reaction thus can yield products of NiAl banded (for low T_o) or multiple phases, $e.g.$, NiAl, Ni_2Al_3 and $NiAl_3$, (for high T_o, slow cooling and higher T_c). Other solidification products could be seen if cooled much slower or the synthesis begun with a higher initial temperature, T_o, which normally leads to a very high combustion temperature (T_c), full mixing and then the precipitation of equilibrium compounds during cooling, e.g.,

$$3Al(Ni,Fe)_{(l)} \rightarrow [xNiAl(Fe)_{(s)}+yNi_3Al[Fe]_{(s)}+z[Al[Ni,Fe]_{eutctic}]+5/2Al(Fe)_{(l)} \quad (3e')$$

The phases that have been reported in the Ni-Al systems are shown in Table 1 [10,12,29,42-47]. An important point to note is that significant variations in the final product chemistry are possible depending on the *initial* processing conditions chosen. It may be noted from Table 1 that the only the NiAl phase is observed when the equimolar NiAl composition is ignited at a lower T_o temperature (e.g., <400°C) than at when ignited at a higher T_o temperature. An increase in the ignition temperature increases the final combustion temperature and the phases, *such as* Ni_2Al_3 or $NiAl_3$ are noted along with NiAl, in the synthesized object.

Equation 3 indicates that the intermediates are $Ni_{(l)}$, $Ni_2Al_3/NiAl_3(Fe)_{(r,s)}$, and Al liquid of at least two concentrations. The $NiAl(Fe)_{(r,s)}$ initially could be in the disordered state when it forms and then order during cooling. As this phase forms directly from the discrete Ni liquid centers, it may manifest as a nano-band formed in the manner described above. A nano-banded structure comprising the $NiAl_{(r,s)}$ and $NiAl_{(s)}$ thus forms in the main grain, i.e., from a combination of Eq. (3b) and Eq. (3d). Equation (3e') is the higher temperature version of equation (3e), i.e., when T_o (the initial compact temperature) is high. It appears that the bands are not seen with an increase in the preheating temperature T_o (preheating is a change in initial conditions) although the microstructural evidence to state this as a general conclusion yet does not exist to the extent required. When the T_o is very high, the condition approaches melt processing. It should be noted that melt processed NiAl, perhaps one that does not undergo a BZ condition, displays a lower lattice parameter (0.285-0.287 nm) [41,54,60] when compared to micropyretically synthesized solid.

X-RAY ANALYSIS

Several time-resolved X-ray studies have been carried out and reported [24,26,27,29,32,46,60,61,78-81] which indicate the occurrence of the intermediates during micropyretic synthesis of NiAl compound. The list of reactants, intermediates, and products reported in different studies at different time intervals from time-resolved X-Ray diffraction patterns are shown in Table 2 [4,26,27,29,46,60,61,80,81]. Wong et al. [60] were the first to report real-time synchrotron diffraction data to monitor the phase transformation in a NiAl micropyretic reaction on a sub-second time scale level less than 100 milliseconds. More recently to the measurements have been made within 10 milliseconds time-frames [46,60,80]. These X-ray results on Ni-Al show that (1) intermediary phases form in the first 0.1 s of synthesis and that the initial NiAl that forms is disordered and only later transforms during cooling to a fully ordered B2 structure [46,60]; (2) a cycle of intermediary formation and dissolution of intermediate Ni-Al compounds is noted particularly that of Ni_2Al_3 [24,46,60-61]; (3) the early stage diffraction indicates extremely small grains (grainy patterns) with no particular orientation as would be noted if nano-crystals are formed [32]; (4) the NiAl diffraction peak-widths narrow with time [32,46,60]; and finally (5) the diffraction patterns indicate ordering, graininess and texturing [32,60] as would be noted from the expected grain growth and ordering in thermally constrained samples. When coupled with the microstructure features presented in this article, it appears that the main hypothesis of this article could be valid i.e., that the micro-kinetic steps could involve BZ reactions in NiAl synthesis. The intermediates have been identified often as at least $NiAl_3$, and Ni_2Al_3 [24,29,32,46,82]. The asymmetric rise and fall of intermediaries which has been reported [24,46,60,61] is also one feature commonly noted in the concentration-time plots of the intermediaries in typical BZ reaction plots [1,31]. Some authors [27,87] have noted that the background intensities in time-lapse continuous X-Ray diffraction patterns studies are related to non-crystalline phases such as molten aluminum, and have noted that this noise oscillates with time. The appearance and disappearance of intermetallic phases in the Ni-Al alloy system has also been reported in references 24, 26, 27, 29, 46, 60, and 80. The disappearance of the reactants, namely Ni and Al, are noted first and subsequently the occurrence of changing amounts of intermediates, including $NiAl_3$, Ni_2Al_3, and NiAl, has been reported (see Table 2). The aluminum and nickel lines disappear on melting. However, the diffraction background level noticeably is found to rise as Al melts, due to the contribution of the amorphous phase [27,80]. This noise level fluctuates during the process. The corresponding thermal analysis [10] shows that the temperature, which corresponds to the diffraction pattern at the time of the formation of intermediates (0.135 s) in reference 27, is recorded at temperatures even as high as $1540°C$, i.e., considerably higher than the melting point of Al ($660°C$). For the BZ reaction to occur, liquid Al must disappear and reappear even with increasing concentration which can eventually change the type of bifurcation (Eq. (2)). Unfortunately there can be no diffraction evidence for this except as noted above from a changing background level and from thermal oscillation results discussed below. In more recent studies on Ni-Al alloy synthesis made with extremely high resolution time lapse X-Ray machines (<10 ms resolution) evidence for very small nano-particle formation (with no orientation features) and with diffuse B2 crystal structure [78,79] have been reported. With time, the patterns evolve into a more textured display (with grain coarsening). The reason for the texturing is not entirely clear [78,79] and could be from band orientation that develop from the heat flow directionality i.e., an interaction of micro and macro-kinetics, or from rapid coarsening.

Table 2. The time range for reactants, intermediates, and products reported in time-resolved X-ray studies. Both NiAl and NiAl mixtures are analyzed along with other micropyretic systems for comparison [26,27,29,46,60,61,80,81].

Components	Reactants Time range(s)	Intermediate, Time range(s)	Products, Time -range(s)	Ref.
Ti:C:Ni:Al= 1:1:1:1	Al: 0.0 s -1.4s Ni: 0.0 s -1.5 s Ti: 0.0 s -1.6 s C: 0.0 s -1.9 s	Several 1.4 s- 1.8s	TiC: 1.4 s - end NiAl: 4.2 s – end	26
Ni:Al=1:1	Ni: 0.0 s -0.135 s Al: 0.0 s -0.135 s	0.135 s -0.345 s (background noise changing due to the melting of Al)	NiAl: 0.135s - end $NiAl_3$: 0.345s - end Ni_2Al_3: 11.205s - end Ni_5Al_3: end	27
Ni:Al=1:1	Ni / Al : 0.0 s – 35.8 s	$NiAl_3$:29.1 s – 39.2 s	NiAl: 33.3 s - end	29
Ni:Al=1:1	Ni: 0.0 s -0.1 s Al: 0.0 s -0.1 s	0.1 s - 10.3 s 31.3 s – 64.3 s	NiAl: 10.3 s - 31.3 s NiAl: 64.3 s – 190.3 s	60
Ni:Al=1:1	Ni :0.0 s -.14.56 s Al :0.0 s -.8.96 s	Ni_2Al_3 : ~14.56 s melting liquid : ~14.56 s	NiAl: end	46
Ti:C:B=2:1:2	Ti : 0.0 s – 8.7 s	No crystalline phase is detected	TiC: 8.3 s – end TiB_2: 8.3 s - end	61
Ta:C=1:1	Ta : 0.0 s – 6.7 s	Ta_2C : ~7.5 s	TaC: ~6.7 s TaC: 30.0 s - end	80
Cu:Al=3:1	Cu: 0.0 s -0.4 s Al: 0.0 s -0.3 s	No crystalline phase is detected	Cu_9Al_4: 0.4 s – end Cu_3Al : 1.6 s - end α Cu : 0.4 s - end	81

THERMAL PROFILE ANALYSIS

Temperature plateaus are commonly noted features in combustion thermal profiles. Plateaus are promoted by lower particle size, solute additions and high initial temperatures, as shown as shown in Figs. 2(a) and 2(b). Such plateaus are a feature noted in almost all thermal studies made on NiAl and other micropyretic systems in references 25, 41, 62, 63, 64, and 90. When such plateaus are noted, the small oscillations that one would expect to see from BZ oscillations that have an endothermic character often become unmasked. Typically these oscillations represent the exothermic and endothermic or heat loss features of the sub-reactions in Eq. 3 (for example Eq. (3b) could be exothermic and Eq. (3d) could be endothermic for a particular formulation or alloy system). The temperature positioning of the plateaus has never been adequately explained in the literature. Typically the plateaus temperature range from the melting point of nickel to the highest melting point of the equimolar NiAl compound. The temperature of the plateau depends on the volume fraction of the Ni band (Eq. (3b)). An increase in the Ni bands reduces the exothermic heat of the sub reaction Eq. (3b), however, the exothermic heat of the subsequent reaction (Eq. (3c)) is correspondingly increased. Equations (3b) and (3c) happen over time, further leading to the cyclic reactions in a volume space.

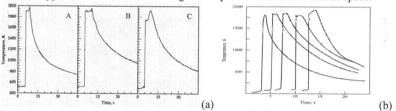

Figure 2. (a) The thermal history for the Ni+50 at. % Al reaction for different Ni particle sizes in the self-propagating mode of micropyretic reaction. The average Ni particle size from left to right is 2.7 μ m, 7.6 μ m, and 27.3 μ m. The compaction pressure is 103 MPa and the initial temperature is 523 K [90]. (b) The thermal history for the different alloys (1) NiAl, (2) $Ni_{49}Al_{49}Ti_2$, (3) $Ni_{48.5}Al_{48.5}Ti_3$, (4) $Ni_{46}Al_{46}Ti_8$, (5) $Ni_{45}Al_{45}Ti_8B_2$ are shown The particle sizes of Al and Ni are 44 μm (-325 mesh) and the initial temperature ranges from 473 K to 623 K [41]. Note variations in the plateau temperature, oscillations in the plateau region and endothermic behavior in the plateau region.

NANO-PARTICLES AND NANO-BANDS

For the high toughness NiAl dilute alloys, extremely tiny, ~50-100 nm size particles in the Fe and V containing NiAl alloys have been seen in the microstructure along with the nano-bands [31]. Typical microstructures are shown in Figs. 3(a)-3(c). Thin bands have also been reported by Curfs et al. [26,27] but without any corresponding report on mechanical properties. The addition of Fe increases the combustion temperature (by about 50K) as do Ti, Cr and Nb additions when compared to the unalloyed equimolar NiAl alloys [40,41] for the same conditions of combustion initiation. This is a surprising result because the exothermic heat release is reduced by additions of solutes such as Fe. The increase in combustion temperature correlates with the increase in ductility but not the toughness. When only Ti is added, the combustion temperature falls with increasing Ti addition to NiAl, but the combustion velocity increases [41]. When measured carefully, thermal oscillations, close to peak temperature are always noted [25,41,46] for all the NiAl alloys.

The nano-band fault regions that are noted within the NiAl grains for several dilute alloys, was first reported in references 31 and 40. In the $Ni_{49.25}Al_{49.5}Nb_{0.75}Ti_{0.5}$ alloy (Fig. 3(d) [40]) weak beam electron imaging showed high and low dislocation concentrations in alternating bands. The fault band spacing decreases from about 100nm in unalloyed NiAl [64] to the very fine 10 – 50nm bands with alloying additions (Fig. 3(d)) [40]. The extensive TEM analysis reported in references 31, 40, and 41 showed that the complex dislocation pattern was unaltered in NiAl alloys containing small amounts of Fe, V or Nb. An examination of the dislocations and chemistry at the grain boundaries indicated similarity with and without the minor solute additions except for the appearance of the nano-particles and nano-bands with the addition of solutes such as Fe, Cr, V [31,41]. No grain boundary segregation was noted with the solute additions except when the alloying content was increased well beyond the solid solubility conditions. The small additions of Fe and other solutes led to increases in the mechanical properties. A detailed examination of the NiAl microstructure with and without the Fe, did not show any <111> slip in NiAl, nor were there any other indication of any other significant microstructural differences noted from the 4 fracture surfaces. Fracture was transgranular in single phase alloys and only became intergranular after the Nb and Ti was increased beyond a total 10 at.% [40]. Similar results have been reported for micropyretically synthesized FeAlMn alloys [33].

Figure 3 Grouping of microstructures which show faults (nano-bands) and nano-particles that are formed during micropyretic synthesis of NiAl (from low to high magnification). (a) Microstructure at low magnification showing the NiAl dendrite formation in alloy $Ni_{50}Al_{48.5}Cr_{0.5}Fe_{0.5}V_{0.5}$ [31]. (b) Intra-grain band structures in the alloy $Ni_{50}Al_{48.5}Cr_{0.5}Fe_{0.5}V_{0.5}$ [31]. (c) Bright-field electron micrograph showing V_2NiAl nano-particles noted at the low angle fault boundaries ($Ni_{50}Al_{48.5}Cr_{0.5}Fe_{0.5}V_{0.5}$) [31]. (d) TEM weak beam image showing <100> type dislocation in alternating bands of the micropyretically synthesized $Ni_{49.25}Al_{49.5}Nb_{0.75}Ti_{0.5}$alloy [40].(e) TEM micrograph showing the nano-band faults noted in the synthesized NiAl containing Ti [41]'

PROCESSING ROUTES FOR KINETIC MANIPULATION.

If the main hypothesis of this article connecting BZ and nano structure formation is valid then one of the key processing conditions for obtaining BZ waves is to hold k_{3f} small enough while increasing k_{2f} in order to enable the nano-structure forming bifurcations. Future studies will explore this as a process control tool. It is known that low temperature chemical waves of Belousov-Zhabotinsky reaction type, can be frozen-in by precipitation of intermediates as shown in references 7 and 8. The distribution and size of the precipitates corresponds to the wave structure and length of propagating chemical waves. The affinity (A) of chemical reactions for low temperatures BZ is such that ($-A$) << $k_B T$ (k_B is the Boltzmann constant). Such an inequality allows for a more tractable determination [8] of the rate of entropy production, σ, where:

$$\sigma = -\sum A(j)/T * w(j) \qquad (4)$$

A and w are the affinities and reaction rate per unit volume for the reaction j. The rate of entropy production difference between a stable periodic trajectory and unstable steady state is given as:

$$\Psi(\alpha_0) = \sigma(\alpha_0) - \sigma(x^*(\alpha_0)) \qquad (5)$$

Where α_0 is a scalar control parameter such as the initial temperature T_0, and x^* is a state of the dynamical system. $\Psi(\alpha_0)$ has a non linear positive value for several combinations of initial conditions and hence allows for bands and precipitates (see below for size determination) regardless of the endothermic possibility in Eq. (3d). For a spatio-temporal BZ it appears that there is no one governing rate production formulation (such as minimum entropy production hypothesis) which can be used to predict the BZ, yet the possibility of $\Psi(\alpha_0)$ positive can possibly be met over a wide variety of conditions. The calculation for $(-A) \gg k_B T$ is ongoing, but there is no reason to doubt that a similar positive entropy rate production is possible. The key issue regardless is the fact that nano-band faults stabilize because of wave particulate interactions. In low temperature systems the precipitates formed for example by the addition of silver ions to a gel layer [7,76], in which chemical waves of the Belousov-Zhabotinsky reaction propagate, are not distributed homogeneously but are stable. The distribution is determined by the shape of the waves at the moment of interaction. Chemical waves are conserved as a periodic pattern by stopping the oscillating chemical process e.g., by the high concentration of added silver ions [7]. The distribution of the precipitates in the NiAl can be interpreted as the precipitation of an intermediate species from the oscillating chemical reaction and appear to have been captured by the X-Ray studies discussed above.

The reaction rate constants of the sub-reaction (Eq. (3b)), namely (k_{2f}) and overall reaction (k_f) as well as the propagation velocity (Vc) may be modeled and calculated from an overall reaction formulation described. Such a modeling method is referred to as a Merzhanov calculation [62]. Solutions are numerically obtained. The combustion temperatures, propagation velocities, and lengths of reaction zone are calculated for each different initial temperature in the numerical modeling. The rate of overall reaction (k_f) is the slowest one in the sub-reaction and is computed from the calculated propagation velocity of the combustion front and the length of reaction zone. Since k_{2f} is proportional to $\exp(-\Delta G/RT)$, the k_{2f} can be determined from the ΔG (change in Gibbs free energy) for a reaction and the calculated combustion temperature. . Since the reaction rate constants are dependent on temperature, the k_{2f} and k_f are respectively calculated to be 4.9×10^6 s^{-1} and 5.0×10^3 s^{-1} at 300K and 5.2×10^6 s^{-1} and 5.8×10^3 s^{-1} at 700K respectively from a Merzhanov calculation for NiAl. A plot of the k_f's with temperature is shown in Fig. 4. Our calculations show that the propagation velocity increases from 583 to 1167 mm/s with an increase in the temperature.

Because of the extremely high frequency of oscillations, i.e. for Eqs. (3b, 3c) and (3d), one can establish the order of magnitude grain diameter (d) and nano-structure size (npd) by combining information from macro- and micro- kinetics. Typical velocities (V_c) of the overall transformation calculated from the macro-kinetics formulations are 0.1 m/s for NiAl [31,40,46]. The overall k_f (for the sequence which most likely will be set by k_{5f} and k_{2f}) are calculated in reference 1 for typical micropyretic reaction sequences. The 'd' is therefore in the order of V_c/k_f and 'npd' is of the order of V_{BZ}/k_{2f} (V_c is the velocity of the combustion front and V_{BZ} is the velocity of the BZ zone movement.). Assuming $k_f = 10^3$ s^{-1} and $V = 0.10$ m/s yields a grain size of the order of a tenth of a millimeter. With $k_{2f} = 10^5$ s^{-1} (obtained from micro-kinetics [1]) and $V_{BZ} \sim 0.01$ m/s yields a 100 nm nano-particle size and a nano-band fault size. The oscillations for the low and high k values are shown in Fig. 5. Note from Figs. 5(a) and (b) that the time period for the intermediate formation is 7.5 s (Figs. 5(a) and (b)). This period is decreased to 0.0075s as the reaction rate constant k is increased to 1000 (Fig. 5(d)). Also shown are the two length scales of bands that have been observed in the low and high temperature reactions. Micro-scale bands from reference 7 for the silial gel silver nitrate solution and nano-bands in NiAl from reference 19 are shown in Fig. 5(c). The reaction rate constant k_{2f} has to be maintained high enough and k_{4f} has to be low so that the bifurcation, i.e., Eq. (2), is enabled. In principle it may be possible to control these values with changes in initial conditions in order to create a desired nano-scale microstructure. Regardless, if the temperature is too high, coarsening of the very fine particles can also occur which disturbs a possible BZ signature analysis after the synthesized object returns to room temperature. This could explain why particles and intra-grain bands are only noted for a limited initial temperature range for the NiAl micropyretic reaction synthesis [31,40,41].

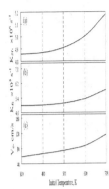

Figure 4. The velocity and temperature of the combustion front as a function of the initial compact temperature can be calculated by a Merzhanov type formulation with an overall heat release rate for the overall reaction and one activation energy. This information can in turn be used to infer reaction constants. The plot shows variations of (a) the reaction rate constant, k_{2f}, of the sub-reaction (Eq. 3(b)), and (b) the overall reaction rate constant (k_f) which will be equal to the slowest sub reaction, and (c) the propagation velocity (V_c); as a function of T_o, the ambient temperature of the unreacted powder mixture. The equimolar NiAl composition with 6 at% NiAl diluent is used as an example for the calculations. The reaction rate constant in sub reaction (2), (k_{2f}), is calculated as exponential ($K_o \times$exp($-\Delta G/RT$)) where K_o is taken as 2×10^8 1/s. ΔG is available from standard reaction tables. The cooling rate during solidification (sub Eq. (3e)) is the slowest reaction rate among all the sub reactions and this rate determines the overall reaction rate constant. With an increase in T_o, the combustion temperature and velocity of the propagation front also increase. The nano-band and nano-particle formation, by the BZ are bound by a lower T_o value which is determined by k_{2f} and k_{3f} in order to allow a bifurcation (Eq. 2)

Regardless of the various observations, correlations, and calculations, offered in this article to describe nano-scale microstructural formation by a high temperature BZ reaction, we recognize that there is no one direct *insitu* measurement which offers conclusive evidence for the proposed sub-reaction sequences shown in Eq. (3) for a spatio-temporal high frequency BZ. The problem stems from the spatial size of the probes that are feasible for the measurements (typically in the order of 100 microns).

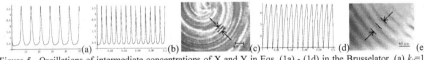

Figure 5. Oscillations of intermediate concentrations of X and Y in Eqs. (1a) - (1d) in the Brusselator. (a) k_{if}=1.0, intermediate=X, (b) k_{if}=1.0, intermediate=Y, (d) k_{if}=1000, intermediate=X,. The vertical and horizontal axes are the dimensionless concentrations of the intermediates and time(s), respectively. The corresponding micron-level and nano-level bands are shown in (c) and (e) respectively [7,40].

CONCLUSIONS

The formation of NiAl during micropyretic synthesis may occur in a regime which involves several sub-reactions [1,9]. Depending on the initial conditions chosen, including the initial alloy chemistry or initial temperature or initial power, the sub-reactions could display BZ type oscillations. Such oscillations will lead to the dispersion of nano-particles and nano-size-banded-faults in the bulk alloy. Uncommon microstructural features such as 100nm nano-bands and nano-particles from such reactions are presented in this article. Supporting thermal analysis and time resolved synchronous X-ray diffraction reports are analyzed. An analysis of quenched microstructures is also presented. All results point to the possibility of the oscillatory BZ reaction. The BZ reaction is modeled as a simple Brusselator which yields the necessary information on the frequency and length scale of the product microstructure by coupling the macro- and micro-kinetics of formation.

The recognition of the nano-particles and nano-bands in high temperature micropyretic reactions can give rise to unique microstructural possibilities. The nano-scale microstructural features are supported by reaction constants exceeding 10^3 s^{-1}. Such high reaction constants are feasible in the high temperature micropyretic reactions. A comprehensive comparison is made for all microstructural and reported property measurements with and without invoking BZ reactions for NiAl. This comparison has indicated that the BZ is a reasonable postulate.

Our continuing effort is to examine issues related to entropy rate and total entropy generation in order to explore additional pathways for synthesis.

REFERENCES

1. Li, HP. & Sekhar, JA. *Acta Mater.*, **57**, 5430-5444 (2009).
2. Epestein, IR. & Pojman, JA. Oxford University Press, Oxford, 1998.
3. Zhabotinsky, AM. *Biophysica*, **9**, 306-311 (1964).
4. Maeda, S. Hara, Y. Yoshida, R. & Hashimoto, S. *Angewandte Chemie International Edition*, **47(35)**, 6690-6693 (2008).
5. Motoike, I. & Adamatzky, A. *Chaos, Solitons & Fractals*, **24**, 107-114 (2005).
6. Kondepudi, D. Introduction to Modern Thermodynamics. Wiley, 2008.
7. Köehler, JM. & Müeller, SC. *J. Phys. Chem.*, **99**, 980-983 (1995).
8. Nicolis, G. & Prigogine, I. Self Organization in Non equilibrium Systems. John Wiley and Sons, New York, 1977.
9. Li, HP. & Sekhar, JA. *International Journal of Self-propagating High-Temperature Synthesis*, **18(4)**, 219-234 (2009).
10. Chen, T. Hampikian, JM. & Thadhani, NN. *Acta Mater.*, **47(8)**, 2567-2579 (1999).
11. Meyers, MA. Mishra, A. & Benson, DJ. *Prog. Mater. Sci.*, **51(4)**, 427-556 (2006).
12. Lucadamo, G. Yang, NYC. San Marchi, C. & Lavernia, EJ. in MRS proceedings, proceeding EE5.2, Ed. By Rudd, RE. Balk, TJ. Windl, W. & Bernstein, N. (2005).
13. Ardell, AJ. Kim, D. & Ozolins, V. *Inter. J. Mater. Res.*, **97(3)**, 295-302 (2006).
14. Lagos, MA. Agote, I. Gutierrez, M. Sargsyan, A. & Pambaguian, L. *International Journal of Self-Propagating High-Temperature Synthesis*, **19(1)**, 23-27 (2010).
15. Dey, GK. Biswas, A. Roy, SK. & Banerjee, S. intermetallic compound. *Materials Science and Engineering A*, **304-306(1-2)**, 641-645 (2001).
16. Zhang, J. Huang, F. & Lin, Z. *Nanoscale*, **2**, 18-34 (2010).
17. He, G. Eckert, J. Löser, W. & Schultz, L. *Nature Materials*, **2(1)**, 33-37 (2003).
18. Barnard, AS. *Nanoscale*, **1(1)**, 89-95 (2009).
19. Klabunde, KJ. & Richards, RM. eds. Nanoscale Materials in Chemistry, Second Edition. John Wiley and Sons, New Jersey, 2009.
20. Balogh, LP. *Nanomedicine: Nanotechnology, Biology and Medicine*, **5(1)**, 1 (2009).
21. Lam, CW. James, JT. McCluskey, R. & Hunter, RL. *Toxicol. Sci.*, **77(1)**, 126-134 (2004).
22. Service, RF. *Science*, **300(5617)**, 243 (2003).
23. http ://www.process-heating.com/Articles/Industry_News/BNP_GUID_9-5-2006_A_10000000000000825366
24. Curfs, C. Turrilas, X. Vaughan, GBM. Terry, AE. Kvick, A. & Rodrı´guez, MA. Submitted for Publications. Preprint link https://webmail.csic.es/bigfiles/descarga.php?l=51872744q&t=1247812056&f=AlNi-ILL.pdf.
25. Zhu, P. Li, JCM. & Liu, CT. *Mater. Sci. Eng. A*, **329-331**, 57-68 (2002).
26. Curfs, C. Cano, IG. Vaughan, GBM. Turrillas, X. Kvick, A. & Rodrı´guez, MA. *J. Euro. Ceram. Soc.*, **22**, 1039-1044 (2002).
27. Curfs, C. Turrillas, X. Vaughan, GBM. Terry, AE. Kvick, A. & Rodrı´guez, MA. *Intermetallics*, **15(9)**, 1163-1171 (2007).
28. Sekhar, AS. *Russian Chemical Reviews*, **77(1)**, 21-37 (2008).
29. Rogachev, AS. Gachon, JC. Grigoryan, HE. Illekova´, E. Kochetov, NF. Nosyrev, FN. Sachkova, NV. Schuster, JC. Sharafutdinov, MR. Shkodich, NF. Tolochko, BP. Tsygankov, PA. & Yagubova, IY. *Nuclear Instruments and Methods in Physics Research A*, **575**, 126–129 (2007).
30. Hsiung, LC. & Sheu, HH. *J. Alloys Compounds*, **479**, 314–325 (2009).
31. Dey, GK. & Sekhar, JA. *Metall. Mater. Trans. B*, **28**, 905-918 (1997).
32. Tolochko, BP. Sharafutdinov, MR. Titov, VM. Zhulanov, VV. & Rogachev, AS. Proceeding to appear in *International Journal of SHS* in regular issues. Presentation number ST2-29, (2009).
33. La, P. Bai, Y. Yang, Y. & Zhao, Y. *Adv. Mater.*, **18(6)**, 733-737 (2006).
34. Li, HP. & Sekhar, JA. Processing Maps for the Micropyretic Synthesis of Structural Composites and Intermetallics, in *Proceedings of the First International Conference on Advanced Synthesis of Engineered Materials* edited by Moore, J.J. Lavernia, E.J., & Froes, F.H. 25-31 (1993).
35. Nishiyama, Z. Materensitic Transformations in Metal and alloys. Academic Press, NY, 1978.
36. Taylor, A. & Doyle, NJ. *J. Appl. Cryst.*, **5**, 201-209 (1972).
37. Hase, K. Hoshino, T. & Amano, K. *Kawasaki Steel Giho*, **34(1)**, 1-6 (2002).
38. Valiev, R. *Nature*, **419**, 887 (2002).
39. Wang, YM. Chen, MW. Zhou FH. & Ma, E. *Nature*, **419**, 912 (2002).

40. Dey, GK. & Sekhar, JA. *Metall. Mater. Trans. B*, **30**, 171-188 (1999).
41. Dey, GK. Arya, A. & Sekhar, JA. *J. Mater. Res.*, **15(1)**, 63-75 (2000).
42. Arkens, O. Delaey, L. Tavernier, J. de. Huybrechts, B. Buekenhout, L. & Libouton, JC. *Mat. Res. Soc. Symp*, **133**, 493–498 (1989).
43. Doty, H. & Abbaschian, R. Structural materials: properties, microstructure and processing. *Mater. Sci. Eng.*, **A195**, 101–111 (1995).
44. Morsi, K. McShane, H. & McLean, M. Formation of Ni3Al and NiAl by hot extrusion reaction synthesis (HERS). *Scr. Mater*, **37(11)**, 1839–1842 (1997).
45. Bhaumik, SK. Divakar, C. Rangaraj, L. & Singh, AK. Reaction sintering of NiAl and TiB_2–NiAl composites under pressure. *Mater.Sci. Eng.*, **A257(2)**, 341–348 (1998).
46. Tingaud, D. Stuppfler, L. Paris, S. Vrel, D. Bernard, F. Penot, C. & Nardou, F. Time-resolved X-ray diffraction study of SHS-produced NiAl and NiAl-ZrO_2 composites. *Inter. J. SHS*, **16(1)**, 12-17 (2007).
47. Li, HP. & Sekhar, JA. Numerical analysis for micropyretic synthesis of NiAl intermetallic compound. *J. Mater. Sci.*, **30(18)**, 4628-4636 (1995).
48. Lahrman, DF. Field, RD. & Darolia, R. in *High Temperature Ordered Intermetallic Alloys IV*, Johnson, L.A., Pope, DP. & Stiegler, JO. eds., Materials Research Socity, Pittsburgh, PA, 603, 1991.
49. Field, RD. Lahrman, DF. & Darolia, R. The effect of alloying on slip system in <001> oriented NiAl single crystals. *Acta. Metall. Mater,* **39**, 2961-2969 (1991).
50. Loretto, MH. & Wasilewski, RJ. Slip systems in NiAl single crystals at 300K and 77K. *Phil. Mag.*, **23(186)**, 1311-1328 (1971).
51. Dollar, M. Dymek, S. Hwang, SJ. & Nash, P. The role of microstructure on strength and ductility of hot-extruded mechanically alloyed NiAl. *Metall. Trans. A*, **24(9)**, 1993-2000 (1993).
52. Cammarota, ·GP. & Casagrande, A. Effect of ternary additions of iron on microstructure and microhardness of the intermetallic NiAl in reactive sintering. *J. Alloys Compounds*, **381**, 208-214 (2004).
53. Cottrell, AH. Vacancies in FeAl and NiAl. *Intermetallics*, **5**, 467-469 (1997).
54. Bozzolo, G. Noebe, RD. & Honecy, F. *Intermetallics*, **8**, 7-18 (2000).
55. Dey, GK. & Sekhar, JA. *Trans. Indian Inst. Met.*, **50(1)**, 79-89 (1998).
56. Noebe, RD. Bowman, RR. & Nathal, MV. *Inter. Mater. Reviews*, **38(4)**, 193-232 (1993).
57. Kaysser, WA. Laag, R. Murray, JC. & Petzow, GE. *Inter. J. Powder Metall*, **27**, 43-49 (1991).
58. Darolia, R. Lahrman, D. & Field, R. *Scripta. Metall. Mater.*, **26(7)**, 1007-1012 (1992).
59. Hong, T. & Freeman, AJ. *Physical Review B*, **43(8)**, 6446-6458 (1991).
60. Wong, J. Larson, EM. Holt, JB. Waide, PA. Rupp, B. & Frahm, R. *Science*, **249**, 1406-1409 (1990).
61. Contreras, L. Turrillas, X. Vaughan, GBM. Kvick, A. & Rodrı́guez, MA. *Acta Mater.*, **52**, 4783-4790 (2004).
62. Mukasyana, AS. & Rogachevb, AS. *Progress in Energy and Combustion Science*, **34**, 377–416 (2008).
63. Rogachev, AS. & Baras, F. *Physical Review E*, **79(2)**, 026214_10 (2009).
64. Gennari, S. Anselmi-Tamburini, U. Maglia, F. Spinolo, G & Munir, ZA. *Acta Materialia*, **54**, 2343-2351 (2006).
65. Thiers, L. Mukasyan, AS. &Varma, A. *Combust. & Flame*, **131(1-2)**, 198-209 (2002).
66. Hunt, EM. Plantier, KB. & Pantoya, ML. *Acta Mater.*, **52**, 3183 (2004).
67. Çamurlua, HE. & Magliab, F. *Journal of Alloys and Compounds*, **478(1-2)**, 721-725 (2009).
68. Torres, RD. Lepienski, CM. Moore, JJ. & Reimanis, IE. *Metallurgical and Materials Transactions B*, **40(2)**, 187-195 (2009).
69. Ozdemir, O. Zeytin, S. & Bindal, C. *Wear*, **265(7-8)**, 979-985. (2008).
70. Bhattacharya, AK. Ho, CT. & Sekhar, JA. *J. Mater. Sci. Lett.*, **11(8)**, 475-476. (1992).
71. Li, HP. Bhaduri, SB. & Sekhar, JA. *Metallurgical and Materials Transactions A*, **23(1)**, 251-261(1992).
72. Li, HP. *J. Mater. Sci.*, **10(6)**, 1379-1386 (1995).
73. Gennari, S. Anselmi-Tamburini, U. Maglia, F. Spinolo, G & Munir, ZA. *J. Phys. Chem. B*, **110**, 7144-7152 (2006).
74. Taguchi, E. Sumida, N. Fujita, H. & Kamino, T. *J. Electron Microsc.*, **39**, 164-167 (1990).
75. Merzhanov, AG. & Khaikin, BI. *Prog. Energy Combust. Sci.*, **14**, 1-98 (1988).
76. Anathakrishna, G. *Physics Reports*, **440**, 113-259 (2007).
77. Li, HP. & Sekhar, JA. *J. Mater. Res.*, **8(10)**, 2515-2523 (1993).
78. Kim, HY. Chung, DS. & Hong, SH. *Mater. Sci. Eng. A*, **396**, 376-384 (2005).
79. Jung, SB. Minamino, Y. Yamane, T. & Saji, S. *J. Mater. Sci. Lett.*, **12**, 1684-1686 (1993).
80. Larson, EM. Wing, J. Holt, JB. Waide, PA. Butt, G. & Termonello, LJ. *J. Mater. Res.*, **8(1)**, 1533-1541 (1993).
81. Merzhanov, AG. *J. Mater. Chem*, **14**, 1779-1786 (2004).
82. Zhao, JF. Unuvar, C. Anselmi-Tamburini, U. & Munir, ZA. *Acta Materialia*, **55**, 5592-5600 (2007).

PHOTOINDUCED SHAPE EVOLUTION OF SILVER NANOPARTICLES: FROM NANOSPHERES TO HEXAGONAL AND TRIANGULAR NANOPRISMS

Thelma Serrano[1], Idalia Gomez[1], Rafael Colás[2,3]

[1]Facultad de Ciencias Químicas, Universidad Autónoma de Nuevo León.
San Nicolás de los Garza, Nuevo León, México
[2]Facultad de Ingeniería Mecánica y Eléctrica, Universidad Autónoma de Nuevo León.
San Nicolás de los Garza, Nuevo León México
[3] Centro de Innovación, Investigación y Desarrollo en Ingeniería y Tecnología, Universidad Autónoma de Nuevo León

ABSTRACT:

This work reports the change in shape from spheres into disks and prisms of silver nanoparticles. The conversion was possible by controlling the concentration of trisodium citrate during a photoconversion process that occurred with different concentration of of Ag:citrate ratio (namely 1:1, 1:2,1:3). The nanoparticles were processed by using magnetic stirring and visible light. It was found that the formation of silver nanodisks was found to be sensitive to the citrate concentration, as it is revealed by UV-Vis spectra. Bigger amounts of citrate produced sharper tips in the prisms; all these effects were witnessed by means of the HRSEM analysis, which showed 30 nm disks and 50 nm prisms.

INTRODUCTION:

Research in nanomaterials has caught the attention of material science as silver and gold nanomaterials offer have interesting chemical and physical properties that can be used in optical, electronic, magnetic and catalytic fields.[1] One of the characteristics of metallic nanoparticles is their anisotropic optical absorption, which is associated to collective oscillations of conduction electrons. This phenomenon is known as surface plasmon resonances (SPR), and it is of high importance in the production of nanophotonic devices and circuits. An important aspect of SPR is its sensitivity to the size and shape of nanoparticles, and these characteristics, as well as their composition, are of paramount importance to determine SPR of metallic nanoparticles. [2]

There is work reported on the change in shape of silver nanoparticles. Thermal [3,4] and photochemical [1,5] treatments are two typical approaches for the morphological modification of silver nanoparticles. Xia et al.[6] reported the formation of triangular silver nanoplates in presence of polyvinyl pyrrolidone (PVP) and sodium citrate; they concluded that light is an essential factor for the generation of triangular seeds, and, in their work, the transformation of spherical colloids into triangular nanoplates took place at elevated temperatures. Mirkin et al.[7-9] demonstrated the growth of silver nanoprisms from spherical silver nanoparticles by photonic energy; they concluded that both, photochemical and thermal control, affect the growth of silver nanocrystals.

The aim of this work is to present a photoinduced method designed to promote the change in shape of silver nanoparticles in the presence of trisodium citrate and visible light at room temperature.

EXPERIMENTAL PROCEDURE

Silver nanoparticles were grown following the procedure describe by...... [ref]. The reactant used were $AgNO_3$ (99.8%), $NaBH_4$ (96.0%) and trisodium citrate (99.0%), which were supplied by CTR Scientific. All chemicals were used without any further purification. The reactants were dissolved in water.

The nanoparticles were prepared by dropping fresh $NaBH_4$ solution (8.0 mM, 1.0 mL) into an aqueous solution of $AgNO_3$ (0.1 mM, 100 mL) that was mixed under magnetic stirring in the presence of three different ratios of trisodium citrate (0.1, 0.2 and 0.3mM). The solution obtained showed a yellowish tone, and was then irradiated with a conventional 70 W sodium lamp during different periods of time, which was the visible light source for the photoinduced process.

Characterization was carried out by UV-Vis analysis performed on a Perkin Elmer UV-Vis Spectrophotometer Lámbda 12. Infrared spectra were obtained with a Perkin Elmer Paragon 1000PC Spectrometer. Images of the nanoparticles were obtained in an high resolution scaning electron microscope (HRSEM) Hitachi S5500 in STEM mode.

RESULTS AND DISCUSSION

A series of color changes were observed during the course of the reaction. The time for the change in color in the different trials was related to the citrate concentration in the mixture. It was observed that the fresh solution was tinted in a yellow tone, and it turned to green after being irradiated for more than 3 h, the tone changed to blue after 36 h of exposure to visible light. The solution prepared with an Ag:citrate ratio of 1:1 only showed the change in tone from yellow to light green; the mixture prepared with the Ag:citrate ratio of 1:2 showed changes from yellow to dark green; the experiment in which a 1:3 ratio of Ag:citrate was used was the only one to achieve the blue coloration (Figure 1). Usually the yellow coloration is related to the presence of low quality silver nanospheres. Blue coloration is related to the presence of silver triangular nanoprism, and the green coloration is observed in between both stages.[9]

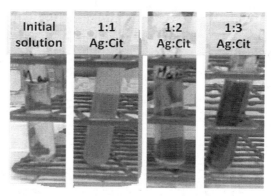

Figure 1. Colorations achieved at the end of irradiation of the samples prepared with different concentrations of citrate.

Figure 2a shows the UV-vis spectra of samples prepared with a 1:1 Ag:citrate ratio and being irradiated for 0, 8 and 24 h. The signal at 770 nm, which can only be detected after an irradiation for 24 h, indicates the formation of nanoprisms. This signal is related to the in-plane dipole resonance, which is very sensitive to the sharpness of the tips of the triangles, as it has been observed that the long-wavelength resonance shifts to 670 nm (without changing the other resonances signals) when the tips of a prisms are rounded[9]. It can also be observed in figures 2, 5 and 8, that the peaks of the in-plane dipole resonance of samples irradiated for 24 hours move from 741 to 779 and 800 nm as the proportion of the Ag:citrate ratio increasses. It can be cosidered that a red-shift of the in-plane dipole resonance peak implies the increase in the size and the aspect ratios of nanoplates, as the in-plane dipole resonance peak is very sensitive to the size and aspect ratio of nanoparticles[9].

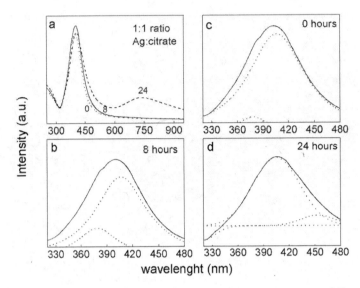

Figure 2. UV-Vis spectra for 1:1 Ag:citrate experiment a) 0, 8 and 24 hours full spectra b) 0 hours from 300-500nm c) 8 hours from 300-500nm d) 24 hours from 300-500nm.

Some interesting features can be observed in the spectra from the 1:1 Ag:citrate experiment in the region from 300 nm to 500 nm. Figure 2b shows the UV-Vis spectrum at time of 0 h; the presence of two peaks were observed at 378 (weak) and 400 (strong). These signals are related to the in-plane quadrupole nanodisk resonance and the presence of spherical silver nanoparticles [5]. Figure 2c shows the UV-Vis spectrum after 8 h of irradiation; the signal at 378 nm increases and the signal at 400 nm decreases in intensity compared to 0 h experiment. The spectrum from the sample irradiated for 24 h shows the presence of peaks at 338 (weak), 400 and 454 (weak) nm. The 400 nm peak is related to spherical silver nanoparticles and it decreases in intensity compared to the spectra from samples irradiated for shorter periods of time. The peak at 338 nm involves two signals: the in-plane dipole resonance for nanodisks and the in-plane quadrupole resonance for triangular nanoprisms. Finally the 454 nm signal involves also two signals: the out of plane quadrupole resonance of nanodisks and the out of plane quadrupole resonance of triangular nanoprism [5,8,9].

All these observations allows for the proposal of a photoconversion mechanism for spherical nanoparticles into triangular nanoprisms. The spherical nanoparticles have a first stage conversion from spherical to hexagonal nanodisks; extra nanospheres convert into triangular nanoprisms as irradiation increases. HRSEM images (figure 3) for the 1:1 Ag:citrate experiment at different times show the presence of silver nanoparticles of around 10 nm in diameter in the non-irradiated solution (Figure 3a). The spherical nanoparticles grow to reach 20 nm in diameter and the presence of hexagonal disks of around 30 nm in diameter can be observed to occur after 5 h of irradiation (Figure 3b). Figure 3c shows the presence of hexagonal nanodisks and triangular nanoprisms after 24 h of irradiation; Figure 3d shows the transformation of a hexagonal nanodisk into a triangular nanoprism by the addition of spherical silver nanoparticles. These observations support the results obtained by UV-Vis analysis.

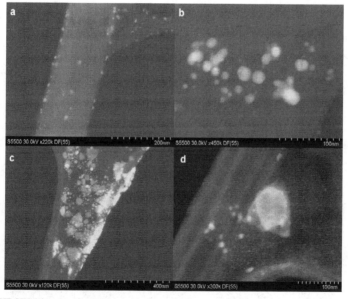

Figure 3. HRSEM images for 1:1 Ag: citrate experiment at 0 (a), 5 (b) and 24 (c and d) hours. Spherical nanoparticles are observed in (a) that transform into nanodiscks in (b); (c) and (d) show the conversion into nanoprims.

Figure 4 shows the UV-Vis spectra for the 1:2 Ag: citrate experiment. At time 0 h a single signal that can be related to the presence of silver nanospheres was found. The spectrum taken after 4 h of irradiation showed a peak at around 416 nm which was composed of peaks at 380 (weak) and 400 (strong) nm that are related to the in-plane quadrupole nanodisk resonance and the presence of spherical silver nanoparticles. After 8 h of irradiation another spectrum was taken. In this case a signal at 410 nm, which can be divided into the signal at 400 nm related to the silver nanospheres, and the weak signal at 332 nm, which was related to the out of plane quadrupole resonance of triangular nanoprism, was observed. The presence of a small peak at 785 nm can be observed, and this one is related to the out of plane quadrupole resonance of triangular nanoprisms. The spectrum of the sample irradiated for 24 h shows the presence of three peaks at 330 (weak) 400 (medium) and 787 (strong and broad) nm. These peaks are related to the out of plane quadrupole resonance of triangular nanoprisms, silver nanospheres signal and the in-plane dipole resonance for triangular nanoprisms. The broadening of the signal at 787 nm gives an idea of the irregular shape at the tips of the triangles and also indicates the existence of larger prisms or prisms aggregation, which induce a strong coupling between individual prisms.

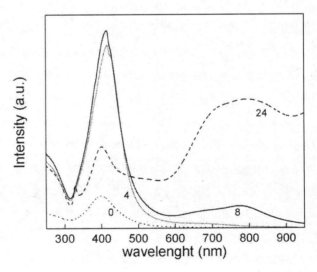

Figure 4. UV- Vis spectra for the 1:2 Ag: citrate experiment at 0, 4, 8, and 24 hours of irradiation.

Figure 5 shows the HRSEM images for the 1:2 Ag: citrate experiment, where at time 0 h (Figure 5a) spherical silver nanoparticles of around 20 nm can be observed. After 4 hours of irradiation (Figure 5b) some triangles and irregular shapes are observed; it was not possible to detect the presence of hexagonal nanodisks. Addition of irregular shape fragments cause the formation of triangular nanoprisms with round tips (Figure 5c). The silver atoms at the corner areas have lower coordination number than those on the surface plane. This results in higher surface energy at the corner areas. Figure 5d shows the aspect of a single silver triangular nanoprism after 24 h of irradiation. In this case, the sides are more stable than the tips, and the absorption of silver atoms on the sides can be favored to achieve sharp tips.

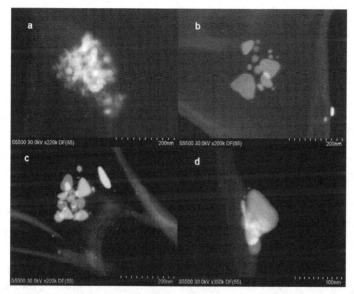

Figure 5. HRSEM images for 1:2 Ag: citrate experiment at a) 0 hours with the presence of spherical silver nanoparticles b) 4 hours with the presence of irregular shape structures that add to get c) silver nanoprism with rounded tips after 24 hours of irradiation d) single triangular nanoprism obtained after 24 hours of irradiation to show the final morphology adopted by the particles on this experiment.

Figure 6 shows the UV-Vis spectra for the 1:3 Ag: citrate experiment. Just like in the 1:2 Ag:citrate experiment, at time 0 h a single signal related to the presence of silver nanospheres was found. After 4 h of irradiation, the spectrum showed a peak of around 410 nm. This peak was composed of two other peaks at 376 (weak) and 400 (strong) nm, which corresponded to in-plane quadrupole nanodisk resonance and the presence of spherical silver nanoparticles. After 8 h of irradiation, another spectrum was observed. In this case a signal at 406 nm, which can be divided into the signal at 400 nm for nanospheres and another weak signal at 331 nm for the out of plane quadrupole resonance of triangular nanoprism is detected. The presence of a peak at 769 nm, related to the out of plane quadrupole resonance of triangular nanoprism, is observed but with a very weak intensity. Three peaks were present in the 24 h spectrum, at 328 (weak), 400 (medium) and 783 (strong) nm. These peaks were related to the out of plane quadrupole resonance of triangular nanoprisms, nanospheres signal and the in-plane dipole resonance for triangular nanoprisms. In this case the signal at 783 nm is better defined than in the experiments with lower concentration of citrate, and it can be due to the change on the tips of the triangles for the final morphology and the formation of better defined nanoprisms.

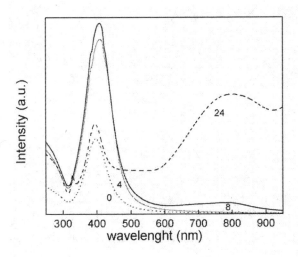

Figure 6. UV-Vis spectra for the 1:3 Ag: citrate experiment at 0, 4, 8, and 24 hours of irradiation.

Figure 7 shows the HRSEM images for the 1:3 Ag: citrate experiment after 24 hours of irradiation. The image shows triangular nanoprism of around 100 nm with 10 nm of thickness that exhibit sharp tips. It is important to notice the large scale conversion obtained in this sample, which was obtained at room temperature by a simple photoconversion process.

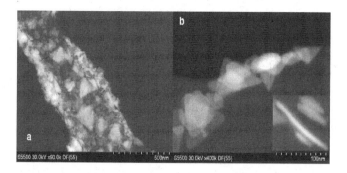

Figure 7. HRSEM images for 1:3 Ag: citrate experiment at with sharp tips after 24 hours of irradiation a) general view of triangular nanoprisms b) the final morphology adopted by the particles on this experiment.

The influence of trisodium citrate on the final morphology adopted by the silver nanoparticles is very important. It determined the final morphology of the nanoparticles and the rate of reaction. It is reported that citrate stabilization is necessary for photoconversion of nanoparticles into prisms. The process was significantly enhanced by increasing the concentration of citrate. The dipole plasmon excitation induced charge separation on the nanodisks and it resulted in the change in shape from disks to prisms.

CONCLUSIONS:

A series of shape conversion from nanospheres into nanodisks and nanoprisms was observed in silver nanoparticles. The concentration of trisodium citrate was crucial for the transformation of nanospheres. Citrate allowed the photochemical process to convert nanospheres to nanodisks through seed-mediated growth. The transformation from hexagons to triangles was observed in the samples prepareerd with a ratio of 1:1 Ag: citrate; higher concentrations inhibited the control. The addition of more nanospheres showed the formation of triangular nanoprisms with modified tips. Higher amounts of citrate reflect in the formation of prisms with sharper edges. The growth process was witnessed by means of UV-Vis spectroscopy and HRSEM. Disks of 30 nm and prisms of 50 nm were obtained.

REFERENCES:

1   Tang B.; Xu S.; An J.; Zhao B.; Xu W. J. Phys. Chem. C, 2009, 113 (17), pp 7025–7030

2   Tang B.; An J.; Zheng X.; Xu S.; Li D.; Zhou J.; Zhao B.; Xu W. J. Phys. Chem. C 2008, 112, 18361–18367

3   Chen, S. H.; Fan, Z. Y.; Carroll, D. L. J. Phys.Chem. B 2002, 106, 10777

4   Wiley, B. J.; Sun, Y.; Xia, Y. Acc. Chem. Res. 2007, 40, 1067.

5   An J.; Tang B.; Ning X.; Zhou J.; Xu S.; Zhao B.; Xu W.; Corredor C.; Lombardi J.R. J. Phys. Chem. C 2007, 111, 18055-18059

6   Sun, Y.; Mayers, B.; Xia, Y. Nano Lett. 2003, 3, 675.

7   Xue, C.; Mirkin, C. A. Angew. Chem., Int. Ed. 2007, 46, 2036.

8   Xue, C.; Met`raux, G. S.; Millstone, J. E.; Mirkin, C. A. J. Am.Chem. Soc. 2008, 130, 8337

9   Jin, R. C.; Cao, Y. W.; Mirkin, C. A.; Kelly, K. L.; Schatz, G. C.; Zheng, J. G. Science 2001, 294, 1901.

SYNTHESIS OF CdS NANOCRYSTALS STABILIZED WITH SODIUM CITRATE

Thelma Serrano[1], Idalia Gómez[1]*, Rafael Colás[2] and José Cavazos
[1]Centro de Laboratorios Especializados, Facultad de Ciencias Químicas, Universidad Autónoma de Nuevo León, México
[2]Facultad de Ingeniería Mecánica y Eléctrica, Universidad Autónoma de Nuevo León, México

ABSTRACT

The results on the study of the synthesis of cadmium sulfide nanoparticles in sodium citrate solutions using microwave radiation are presented. $CdCl_2$ and thioacetamide solutions were prepared with three different concentrations of sodium citrate, 2.0, 1.5 and 1.0 mM respectively, keeping constant a pH of 8, heating was carried out in a microwave oven for 60 s. Synthesis of nanoparticles were confirmed by different experimental techniques. The material was arranged as light sensitive cubic particles of nanometric size. Ultraviolet spectroscopy shows that the energy of the band gap is affected by the size of the particles. It was found that the size and aspect of the particles changed as a function of time when held at 86°C.

INTRODUCTION

Work in nanotechnology allows the modification of already existing materials to achieve characteristics and properties not observed in average-size crystals[1,2]. Synthesis and characterization of semiconductors are of interest to this field because their quantum states depend on the size and self-organization of nanocrystals[3-6].

Cadmium sulfide (CdS) nanoparticles have been obtained by different methods such as synthesis in anionic polymers, membranes, micelles, porous glasses, zeolites, and physical evaporation[7-13]. CdS exhibits interesting chemical, physical and optoelectronic properties and has a high potential use in microelectronic, catalytic and bioanalytical applications, owing to its nonlinear optical and luminescence properties[1-3,7] but most synthesis methods are conducted at relativity high temperatures or use toxic agents such as H_2S or organometallic precursors. We have reported an interesting method to synthesize CdS nanoparticles using simple precursors such as $CdCl_2$ and thioacetamide (CH_3CSNH_2) by microwave heating.[14] The interest in possible uses of CdS in microelectronic systems is the driving force for developing synthesizing methods capable to obtain stable colloidal nanosize particles.

Advances on the control of crystal sizes is required to obtain well defined characteristics and morphologies, some of the mechanisms used to control nanocrystal sizes rely on forcing defined space arrangements[15-18], artificial assembling [15] or preferential nuclei growth [15,19,20]. The methods for stabilization interact with electrostatic forces; chemicals substances used for this purpose affect the coordination reactions and promote growth in different directions by stabilizing free electronic pairs present in the molecular structure[19,20].

The surface modification of the particles can be made by the use of ligands that can introduce a high level of self organization. Some studies on the surface modification of II-VI semiconductors with different organic ligands such as dyes, aromatic hydrocarbons, etc have been performed.[21] Sodium citrate is one of the most common agents used in the synthesis of metallic and semiconductor nanoparticles because it can act as a reducing agent or as a coordination agent. This is due to free electron pairs in the carbonyl group, which stabilize semiconductor nanoparticles by electrostatic forces generated in the system, and it can also act as coordination agent in compounds with metallic atoms with free orbitals.[20,21]

Especially with metals as Cd^{2+} the effect of sodium citrate in the synthesis and size distribution has been analysed. Ortoño report an ammonia-free chemical bath method to deposit highly oriented CdS films on glass substrates. The method is based in the substitution of ammonia by sodium citrate as the complexing agent of cadmium ions in the reaction solution.[22] Also, Li report the synthesis of ultrafine Co nanoparticles with the size of ca. 2 nm. The particles were prepared by an organic colloid method, in which sodium formate as reducing agent, ethylene glycol as solvent, sodium citrate as both complexing agent and stabilizing agent, respectively.[23] And finally Lokhande reports the electrosynthesis of cadmium selenide films from deposition bath containing sodium selenosulphite, along with cadmium complexed with sodium citrate under potentiostatic deposition condition on titanium substrates.[24]

The aim of the present work is to present the results of a study on the influence of sodium citrate concentration and electromagnetic radiation on the morphology adopted by CdS nanoparticles processed by microwave heating.

EXPERIMENTAL PROCEDURE

CdS nanoparticles were prepared with a solution of thioacetamide (CH_3CSNH_2) and of $CdCl_2$ at 30 mM. The Solutions were mixed in a beaker and completed to 50 ml with sodium citrate concentrations of 2.0, 1.5 and 1.0 mM. The pH was adjusted to 8 and the mixtures were heated by microwave radiation for 60 s with 1300 W power at 2.4 GHz to obtain the dispersed particles. The experiments were placed under different radiation conditions. Table 1 shows the experimental conditions.

Table 1. Experimental condition for the different experiments.

Experiment	Sodium citrate concentration	Radiation conditions	Constant Conditions
1	2.0 mM	Natural light	
2	2.0 mM	UV (λ=302nm)	$[S^{2-}] = 30$mM
3	2.0 mM	Darkness	$[Cd^{2+}] = 30$mM
4	1.5 mM	Natural light	pH = 8
5	1.0 mM	Natural light	

The synthesized material was studied with different techniques. The size and dispersion of the particles were determined by ultraviolet-visible (UV-Vis), and infrared (FTIR) spectroscopy and X-ray diffraction (XRD) with copper radiation (λ = 0.15418 nm). Samples were held for varying time at a temperature of 86°C; the change in particle size were monitored with UV-Vis. Synthesized samples were examined via optical microscopy (OM), atomic force microscopy (AFM) and Transmission Electron Microscope (TEM) to analyze the effect of the stabilizer concentration on the morphology of the particles. The microwave system used for the synthesis of CdS NPs operates at 1150W, 2.45 GHz, working at 80% of power under continuous heating. UV-vis absorption spectra were obtained using a Perkin Elmer UV-vis Lambda 12 spectrophotometer. For luminescence quantum yield measurements, a dilute solution of coumarin 1 in ethanol was used as standard. Both the nanoparticle dispersion and the coumarin 1/ethanol solution were adjusted to have an absorbance of 0.10. A corrected luminescence integrated area was used to calculate the quantum yield. Fluorescence experiments were performed using a Perkin Elmer PL Lambda 12 spectrofluorimeter using a wavelength of excitation of 250nm. All optical measurements were performed at room temperature under ambient conditions. Samples were precipitated with ethanol and dried in a vacuum oven for XRD and FTIR characterization. The XRD patterns were obtained from a Siemens D5000 Cu Kα (λ = 1.5418Å) diffractometer. The FTIR spectra were obtained using a Perkin Elmer Paragon 1000. The Optic Microscopy was obtained using an Olympus BX60. AFM images were recorded in a Quesant Q-Scope 3500 atomic force microscope using contact mode, the samples were fixed by solvent evaporation on a metallic plate standard support for AFM microscopy. TEM images were obtained using a Tecnai G2 F20X-Twin DE FEI JEM 2010 JEOL.

RESULTS AND DISCUSSION

Figure 1 shows the X-ray diffraction spectrum of the material processed for experiment 1. The spectrum indicates the peaks corresponding to different reflections of CdS, which has a cubic structure. The size of the particles can be calculated from the broadening of the peaks according to Sherrer equation and it was calculated to be of 15 nm.

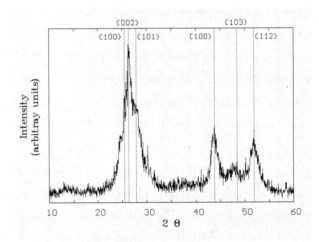

Fig. 1. X-ray diffraction spectrum of samples synthesized with a 2.0mM of sodium citrate.

Nanoparticles of CdS exhibit semiconductor behavior with an energy for the band gap (E_g) of 2.53 eV[25]. This energy can be evaluated from the UV-Vis spectrum by:

$$E_g = h\nu = \frac{hc}{\lambda} = \frac{1240}{\lambda} = \frac{c_1}{r^2} - \frac{c_2}{r} \qquad (2)$$

where h is Planck's constant, c the speed of light and λ the wavelength and c_1 and c_2 are parameters that depend on the experimental conditions. This energy value can be related to the particle size on CdS particles ranging in the size of 2 to 6.5 nm.

Samples synthesized with different concentration of sodium citrate were held at 86°C for varying time; changes in size of the particles were evaluated by means of UV-Vis. Fig. 2 shows the UV-Vis spectra obtained from samples synthesized with 1.0, 1.5 and 2.0 mM concentrations of sodium citrate as a function of time. It can be observed a red shift on the absorption wavelength which is related to an increase of the particle size. Fig. 2d shows the variation in size of the particles as a function of time. Three different stages can be appreciated in the behavior of samples prepared with 1.5 and 2.0 mM concentrations (the samples prepared with the lower concentration were held to shorter times). The first of these stages shows an accelerated growth caused by particles joining together due to electrostatic forces. The second stage is controlled by the concentration of the stabilizer (sodium citrate, in the present case) and is attributed to diffusion. The last stage shows accelerated growth attributed to further joining due to electrostatic forces. The difference in behavior of the curves shown in Fig. 2d is attributed to the concentration of the stabilizer, as it can be appreciated that the slope of the second stage of the sample synthesized with the higher solution is lower than those for the other samples.

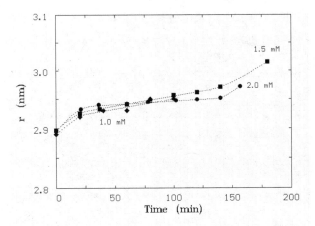

Fig. 2. Changes in the radius of the synthesized particles as a function of time at 86 °C at different concentration of sodium citrate.

A second series of experiments was measured by varying the radiation (natural light, UV and darkness) over the sample and keeping the stabilizer concentration constant to 2.0 mM. Figure 3 shows the variation in size of the particles as a function of time. In the same way that the experiments before, three different stages were appreciated. For natural light, particles were stabilized to size smaller than 6.5 nm for 180min. To investigate the influence of radiation on the behavior of the samples, experiments under UV radiation and darkness condition were investigated. Figure 3b shows the variation in size for samples in darkness conditions. It was observed larger growing rates than the one for samples exposed to natural light. The time to reach similar sizes in both conditions was diminished 30min. Figure 3c shows the variation in size for the experiment under UV radiation (λ= 302nm), in this case particles did not show a significant variation in size for all the exposure time. The difference in behavior of the curves shown in Fig. 3 is attributed to the interaction of samples with light as a photoactive compound. The agglomerates of particles present a fragmentation process due to UV radiation and it stabilizes the particle size. When the sample was kept in darkness, light interactions were not present over the particles and they were free to grow. Finally, as natural light is mainly visible and UV radiation, particles present an intermediate growth behavior.

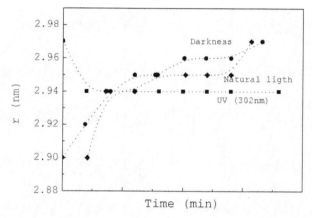

Fig. 3. Changes in the radius of the synthesized particles as a function of time at 86 °C exposed to different electromagnetic radiation.

The morphologic arrangements of the particles were studied by AFM and TEM for the 2.0 and 1.5mM sodium citrate stabilizer concentration. From the TEM images the characteristics observed are the reduction of size and shape of the agglomerates, and morphologies flower like, Fig. 4a, and lineal, Fig. 4b. Fig. 4c. The diagrams in these figures show the results of modeling the arrangement of CdS and sodium citrate, considering that the ligands of Cd(II) and sodium citrate are attached to different species [28]. The last fact base in a FTIR spectrum obtained for the sample where sodium citrate was detected on the surface of the particles. The results modeled with the lower concentration of citrate, Fig. 4b, show that CdS are surrounded by the sodium citrate molecules, promoting the flowerlike arrangement, whereas, when the concentration is increased, Fig. 4c, the arrangements follow a lineal pattern. The particles prepared with the lowest concentration (1.0mM), form arrangements of size around 10 μm with platelets in square and hexagonal shapes. An optic microscopy obtained by transmitted light of them is showed in Figure 4d.

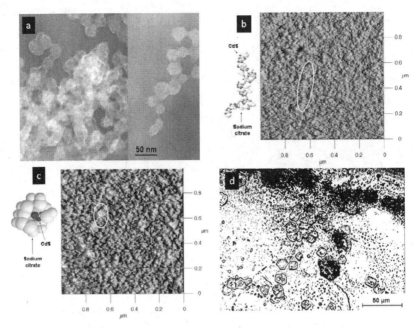

Fig.4. a)TEM images of sample synthesized with a concentration of 1.5mM or 2.0mM of sodium citrate and reheated at 86 °C for 180min. Atomic force microscopy image of a sample synthesized with a concentration of b) 1.5mMof and c) 2.0mM of sodium citrate and reheated at 86 °C for 180 min; the diagram at the left hand side indicates how CdS particles are surrounded by sodium citrate molecules. d) Transmitted light image of sample synthesized with a concentration of 1.0mM of sodium citrate.

For the experiment exposed to natural light and UV radiation, flower like morphologies were observed (similar to previous experiments). In the case of the darkness condition experiment, the particles present rods morphologies with diameters from 50 nm to 1μm (Fig 5). The different morphologies may be influenced by the radiation where light drives the particles to grow into preferential morphologies.

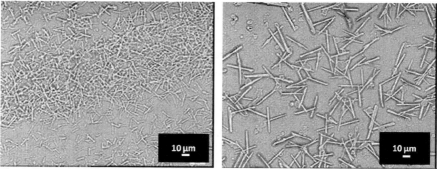

Fig. 5. Transmitted light image of sample synthesized with a concentration of 2.0mM of sodium citrate and kept on darkness conditions.

CONCLUSIONS

Results indicate that CdS can be synthesized into particles smaller than 20 nm in size of cubic zinc blend type structure with mixtures of thioacetamide, cadmium chloride and sodium citrate were heated up by means of microwave radiation.

The growth mechanism may be modified by the concentration of stabilizer and the electromagnetic radiation. The concentration of stabilizer is proportional to the time of stabilization of the particles. The UV radiation breaks agglomerates of particles and it induce the stabilization of the system.

Atomic force microscopy and transmission electron microscopy show that the particles arrange into different types of structures depending on the concentration of sodium citrate. The shape of the arrangements can be explained in terms of the way that sodium citrate molecules surround the particles. In concentration below 1.5mM of sodium citrate, the particles are instable and form agglomerates in the microscopic scale.

REFERENCES

1. X. Xiaoda, M. Stevens, M.B. Cortie, In situ Precipitation of Gold Nanoparticles onto Glass for Potential Architectural Applications, *Chem. Mater.*, 16 (2004) 2259-2266.
2. C. Burda, X. Chen, R. Narayanan, M.A.El-Sayed, Chemistry and Properties of Nanocrystals of Different Shapes, *Chem. Rev.*, 105 (2005) 1025-1102.

3. T. Ni, D.K. Nagesha, J. Robles, N.F. Materer, S. Müssig, N.A. Kotov, CdS Nanoparticles Modified to Chalcogen Sites: New Supramolecular Complexes, Butterfly Bridging and Related Optical Effects, *J. Am. Chem. Soc.*, 124 (2002) 3980-399.

4. M.L. Curri, A. Agostiano, M. Catalano, L. Chiavarone, V. Spagnolo, M. Lugarà, Synthesis and Characterization of CdS Nanoclusters in a Quaternary Microemulsion: the Role of the Cosurfactant. *J. Phys. Chem. B*, 104 (2000) 8391-8397.

5. M. Zhang, M. Drechsler, A.H.E. Müller, Template-Controlled Synthesis of Wire-Like Cadmium Sulfide Nanoparticle Assemblies within Core-Shell Cylindrical Polymer Brushes, *Chem. Mater.*, 16 (2004) 537-543.

6. T. Zhang, W. Dong, M. Keeter-Brewer, S. Konar, R. Njabon, Z.R. Tian, Site Specific Nucleation and Growth Kinetics in Herarchical Nanosyntheses of Branches ZnO Crystallites, *J. Am. Chem. Soc.*, 128 (2006) 10960-10968.

7. Wenzhong W.; Germanenko I.; El.Shall S. Room- Temperature Synthesis and Characterization of Nanocrystalline CdS, ZnS and $Cd_xZn_{1-x}S$. *Chem.Mater*. 14 (2002) 3028-3033.

8. E. Smotkin, M. Brown, L.K. Rabenberg, K. Salomon A.J. Bard, A. Campion, M.A. Fox, T.E. Msllouk, S. Webber J.M. White, Ultrasmall particles of cadmium selenide and cadmium sulfide formed in Nafion by an ion-dilution technique. *J. Phys. Chem.*, 94 (1990) 7543-7549.

9. S. Wang, P. Liu, X. Wang X. Fu, Homogeneously Distributed CdS Nanoparticles in Nafion Membranes: Preparation, Characterization and Photocatalytic Properties, *Langmuir* 21 (2005) 11969-11973.

10. Simmons B.; Li S.; John V.; McPherson G.L.; Bose A.; Zhou W.; He J. Morphology of CdS Nanocrystals Synthesized in a Mixed Surfactant System. *Nano Letters* 2 4 (2002) 263-268.

11. H. Mathieu, T. Rachard, J. Allegre, P. Lefebvre, G. Arnaud, W. Granier, L. Boudes, J. Merc, A. Pradel, M. Ribes, Structure and Properties of the Condensed Phosphates. VII. Hydrolytic Degradation of Pyro- and Tripolyphosphate. *J. Am. Chem. Soc.* 77 (1988) 287-291.

12. N. Herron, Y. Wang, M.M. Eddy, G.D. Stucky, D. Cox, K. Moller, T. Bein, Structure and optical properties of cadmium sulfide superclusters in zeolite hosts. *J. Am. Chem. Soc.*, 111 (1989) 530-540.

13. Y. Wang, G. Meng, L. Zhang, C. Liang, J. Zhang, Catalytic Growth of Large-Scale Single-Crystal CdS Nanowires by Physical Evaporation and Their Photoluminescence, *Chem. Mater.* 14 (2002) 1773-1777.

14. S. Martínez, T. Serrano, I. Gómez, A. Hernández, Síntesis y Caracterización de Nanoparticulas de CdS obtenidas por microondas, *Bol. Soc. Esp. Ceram. V.*, 46 (2007) 97-101.

15. Narayanaswamy A.; Xu Z.; Pradhan N.; Kim N.; Peng X. Formation of Nearly Monodisperse In2O3 Nanodots and Oriented-Attached Nanoflowers: Hydrolysis and Alcoholysis vs Pyrolysis. *J. Am. Chem. Soc.* 128 (2006) 10310-10319.

16. Jie J. S.; Zhang W.J.; Jiang Y.; Meng M.; Li Y.Q.; Lee T. Photoconductive Characteristics of Single-Crystal CdS Nanoribbons. *NanoLetters* 6 9 (2006) 1887-1892.

17. Ghezelbash A.; Koo B.; Korgel B. A. Self-Assembled Stripe Patterns of CdS Nanorods. *NanoLetters*, 6 8 (2006) 1832-1836.

18. Yao W.; Yu S.; Liu S.; Chen J.; Liu X.; Li F. Architectural Control Syntheses of CdS and CdSe Nanoflowers, Branched Nanowires, and Nanotrees via a Solvothermal Approach in a Mixed Solution and Their Photocatalytic

Property. *J. Phys. Chem. B* 110 (2006) 11704-11710.

19. J. Zhu, O. Palchik, S. Chen, A.Gedanken, Microwave Assisted Preparation of CdSe, PbSe and $Cu_{2-x}Se$ Nanoparticles, *J. Phys. Chem.*, 104 (2000) 7344-7347.

20. A.S.C. Samia, J.A. Schlueter, J.S. Jiang, S.D. Bader, C.J. Qin, X.M.Lin, Effect of Ligand-Metal Interactions on the Growth of Transition-Metal and Alloy Nanoparticles, *Chem Mater.*, 16 (2006) 5203-5212.

21. T. Ni, D.K. Nagesha, J. Robles, N. F. Materer, S. Müssig,N.A. Kotov. CdS Nanoparticles Modified to Chalcogen Sites: New Supramolecular Complexes, Butterfly Bridging, and Related Optical Effects. *J. AM. CHEM. SOC.*, 124 (2002) 3980-3992.

22. M.B Ortuño Lòpez., J.J. Valenzuela-Jàuregui , M. Sotelo-Lerma, A. Mendoza-Galvàn, R.   Ramìrez-Bon. Highly oriented CdS films deposited by an ammonia-free chemical bath method. Thin Solid Films 429 (2003) 34–3.

23. H. Li, S. Liao. Organic colloid method to prepare ultrafine cobalt nanoparticles with thesize of 2 nm. Solid State Communications 145 (2008) 118–121.

24. C.D. Lokhande, E.-H. Lee, K.-D. Deog Jung, O.-S. Joo. Electrosynthesis of cadmium selenide films from sodium citrate–selenosulphite bath. Materials Chemistry and Physics 91(2005) 399–40

25. Y. Wang, N. Herron, Nanometer-sized semiconductor clusters: materials synthesis, quantum size effects and photophysical properties, *J. Phys. Chem.*, 95 (1991) 525-532.

26. Nakamoto, K., 1986. Spectra of Inorganic and Coordination Compounds,Infrared and Raman Spectra of Inorganic and Coordination Compounds. Wiley, New York.

27. Sene, C.F.B., McCann, M.C., Wilson, R.H., Grinter, R., 1994. Fouriertransform Raman and Fourier-transform infrared spectroscopy. An investigation of five higher plant cell walls and their components. Plant Physiol. 106, 1623–1631.

28. M. Dakanali, E.T. Kefalas, C.P. Raptopoulou, A. Terzis, T. Mavromoustakos, A. Salifoglou, Synthesis and Spectroscopic and Structural Studies fo a New Cadmium(II)-Citrate Aqueous Complex. Potential Relevance to Cadmium (II)-Citrate Speciation and Links to Cadmium Toxicity, *Inorg. Chem.*, 42 (2003) 2531-2537.

FREEZING BEHAVIOR AND PROPERTIES OF FREEZE CAST KAOLINITE-SILICA POROUS NANOCOMPOSITE

Wenle Li[1], Kathy Lu[1], John Y. Walz[2]
[1]Department of Materials Science and Engineering, [2]Department of Chemical Engineering
Virginia Polytechnic Institute and State University
Blacksburg, Virginia, USA

ABSTRACT

This paper focuses on understanding the kaolinite-to-silica ratio and freezing rate effects on freeze-cast kaolinite-silica nanocomposites. The microstructure evolution, specific surface area, and flexural strength of the resulting composites are examined by scanning electron microscopy, nitrogen adsorption measurement, and equibiaxial strength test, respectively. The experimental results are analyzed from kaolinite-silica gel formation, particle size/shape, and freezing condition points of view. The ice crystal nucleation and growth behaviors during solidification are discussed. The interaction and cooperation between kaolinite platelets and silica nanoparticles during the strength measurement are investigated.

INTRODUCTION

Freeze casting solidifies particle suspensions and sublimates the frozen media under low pressure to fabricate material/composite with tunable shape and controllable microstructure. In the past few years, this technique has been extensively used to fabricate porous ceramics with well-designed pore size and morphology as well as good mechanical properties.[1-6] The solidification behaviors during freeze casting process have been studied.[7-9] Many factors such as additives[3,10-12], freezing temperature[13], freezing rate[14], solidification velocity[7,8], solids loading[15], particle size[16], suspension state[17] can effectively modify the microstructure and the subsequent properties, which make freeze casting a versatile method in creating porous ceramics. Various chemicals are typically used as dispersing media, such as water, camphene, naphthalene-camphor, and tert-butyl alcohol, among which water is a desirable one since its solidification behavior has been well studied[7,8,18]. Both homogeneous[14] and directional[4] microstructures can be obtained using water as the templating media by adjusting the freezing conditions. However, although much effort has been devoted to freeze casting technique and several approaches are developed to modify the properties of resulting materials, a general model for relating the microscopic properties and external conditions is not currently available. The freezing behaviors and properties for different systems need to be separately evaluated.

Recently, a kaolinite-silica system was found to perform a gel transition process when a sufficient amount salt was added.[14,19,20] The plate-like kaolinite particles and silica nanoparticles are able to work cooperatively to increase the strength of the gel.[20] When freeze casting is applied to the gel, a nanocomposite with homogeneous microstructure can be fabricated. This system is interesting and of much value for studying the freezing behavior of ice solidification since its gelled state and bimodal particle size affects the repulsion efficiency of the solidified ice

57

front and subsequently lead to a different mechanism for freeze casting.

The objective of this paper is to understand the kaolinite-silica ratio and freezing rate effects on the microstructure, specific surface area, and flexural strength of the fabricated nanocomposites. The freeze casting is applied by directly exposing samples to low-temperature refrigeration and all the conditions except for the freezing rate are kept the same to eliminate other possible variations. The solids loading effects are investigated by adjusting the kaolinite-to-silica ratio while keeping the total solid concentration fixed, ensuring that the property evolution is induced by the particle interaction. Ice solidification behavior, which is analyzed from nucleation and growth points of view, combined with particle interaction and cooperation, is used to understand the microstructure evolution and strength improvement.

Although not addressed in this paper, our next step will be to investigate the effect of the kaolinite particles on the structure and properties of these composites after sintering at elevated temperature. The ability to control microstructure and properties via the addition of kaolinite could be useful in a variety of applications, such as catalyst supports, filtration media, and structural support elements. The properties measured here (i.e., strength and specific surface area) will be important in assessing the potential use of the composites for these applications.

EXPERIMENTAL PROCEDURE

Raw Materials and Sample Preparation

In this study, 34% wt. silica nanoparticle suspension (Ludox TMA, Sigma Aldrich, St. Louis, MI), sodium chloride (AR grade, Mallinckrodt, Paris, KY), kaolinite (Hydrite Flat-D, Imerys Performance Materials, Dry Branch, GA), and de-ionized water were used to prepare kaolinite-silica gels. The silica nanoparticles had a nominal diameter of 22 nm and a density of 2.37 g/cm^3 [21,22]. The specific surface area of the silica was reported by the manufacturer as ~140 m^2/g, however, calculation using the particle size and density (i.e., $A_{sp} = 3/\rho R$, where ρ is the particle density, R is particle radius, and A_{sp} is the specific surface area) yields a value of approximately 115 m^2/g, which agrees well with BET measurements reported below. The particles were well dispersed in de-ionized water. The kaolinite particles were composed of thin platelets and were polydisperse, with observed diameters ranging between 200 nm and 6 μm and a thickness at 50-200 nm. The density of the kaolinite, obtained by measuring the volume of water displaced by a known mass of added clay, was 2.56 g/cm^3, and the manufacturer reported specific surface area was 7 m^2/g.

Desired amounts of silica nanoparticle suspension and de-ionized water were mixed in a vial using a vortexer (Fisher Scientific Inc, Bohemia, NY). The kaolinite particles were added into the suspension, followed by 1 min mixing. Finally, a specific amount of NaCl solution of 5 M concentration was introduced and the mixture was homogenized for another 1 min (Fig. 1a). A series of samples with kaolinite concentrations of 0% vol., 6% vol., 10% vol., and 14% vol. were prepared. The total solids concentration (kaolinite plus silica) was kept at 18% vol., meaning that as the kaolinite concentration was increased, the silica concentration decreased. The NaCl concentration in all samples was 0.5 M.

After mixing, the suspensions were poured into small aluminum foil containers (8 mm diameter, 3 m m deep, sample volume of approximately 0.15 cm^3) and large silicone rubber containers (31 mm diameter, 2 mm deep, sample volume of approximately 1.5 cm^3) (Fig. 1b) and held there for 4 hrs to allow gelation (Fig. 1c). The small samples were used for microstructure observations and surface area measurements, while the large ones were designed for flexural strength tests. The gelation times of solutions with 0, 6, 10 a nd 14% vol. kaolinite were 30 min, 10 min, 5 min, and > 8 hrs, respectively. (The gelation time is defined here as the time at which the sample would no longer flow upon inverting the sample holder.) The holding time was chosen to be 4 hrs to allow sufficient time for the silica nanoparticles and kaolinite platelets to interact while also minimizing sedimentation. Note that the samples containing 14% vol. kaolinite would typically not gel within the 4-hour holding time, however the sample was sufficiently viscous that little sedimentation of the kaolinte was observed.

After 4 hours, the gels (and containers) were placed into an Advantage Freeze Dryer (SP Industries VirTis, Gardiner, NY) to execute freezing and sublimation (Fig. 1d). The samples were first cooled to 238 K with freezing rates of 2 K/min, 1 K/min, and 0.05 K/min respectively and then held at 238 K and ~1.34 Pa for 400 m in to sublimate the ice. (For purposes of brevity, only the results obtained at 2 K/min and 0.05 K/min are presented here for flexural strength analyses.)

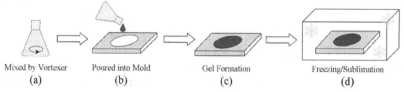

| Mixed by Vortexer | Poured into Mold | Gel Formation | Freezing/Sublimation |
| (a) | (b) | (c) | (d) |

Figure 1. Experimental procedures: (a) suspension preparation, (b) molding, (c) gel formation, (d) freezing and sublimation.

Characterization

The microstructures of the freeze-cast composites were observed using a field emission scanning electron microscope (SEM, LEO1550, Carl Zeiss MicroImaging Inc., Thornwood, NY). The freeze-cast samples were coated with 15-20 nm Au-Pt layer before examination. Images of both the top and interior (cross-section) surfaces of the composite samples were obtained (the cross-sections were obtained by breaking the samples by hand and imaging one of the exposed surfaces). Specific surface area and pore volume/size analyses were carried out by nitrogen adsorption (Autosorb-1 C, Quantachrome Instruments, Boynton Beach, FL). The equibiaxial flexural strength of the freeze-cast composites was measured by a strength test apparatus with a 1 kN load cell (Instron 4204, Instron, Norwood, MA). The test was a t wo-dimensional bending compressive strength measurement with 360° r otation about the load axis, which followed ASTM C 1499.[23] The crosshead was lowered at a speed of 0.1 m m/min until the sample was broken. The equibiaxial flexural strength was calculated as

$$\sigma_f = \frac{3F}{2\pi h^2}\left[(1-v)\frac{D_S^2 - D_L^2}{2D^2} + (1+v)\ln\frac{D_S}{D_L}\right] \tag{1}$$

where σ_f is the equibiaxial flexural strength, F is breaking load, h is specimen thickness, v is Poisson's ratio, D is sample diameter, D_S is support ring diameter, and D_L is load ring diameter.

RESULTS AND DISCUSSION

Ice Solidification Behavior

The ice solidification behavior during the freeze casting process is analyzed to understand the microstructure of freeze cast composites, since the fabricated pores are replica of ice crystals. Among the 15 known crystalline solid phases of water, the hexagonal ice (ice Ih) crystal structure is the most abundant one in nature. Based upon different chemical potentials of each facet, anisotropic growth of ice crystal is kinetically favorable during solidification. However, in this system, it is believed that the anisotropic ice crystal growth is inhibited and the homogeneous microstructure is fabricated due to the low temperature gradient throughout the sample and the high particle hindrance that effectively offset the repulsion of growing ice fronts.

As mentioned above in the experimental procedure section, the thickness of the samples was designed to be very small (2-3 mm) and the studied freezing rates were chosen in the low range (0.05-2 K/min). These conditions work together to ensure a very low temperature gradient throughout the sample, where the directional growth of ice crystals is not favored. In addition, samples were held for 4 hrs before the freeze casting to provide sufficient time either for gel formation (the first three samples) or the enhancement of the viscosity (14% vol. kaolinite sample). The gelled state or high viscosity of the samples largely increases the hindrance force to the growing ice front. In comparison, the driving force of the solidified fronts is not high enough since the temperature gradient is low. Moreover, the large plate-like kaolinite particles form the framework of the sample which further increases the particle hindrance to ice crystal growth (Fig. 2). Therefore, the anisotropic growth of ice crystals is limited and homogeneous ice crystals form, which consequently results in a homogeneous porous microstructure.

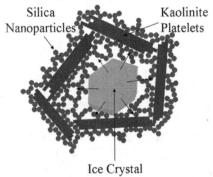

Figure 2. Schematic diagram of ice solidification behavior during freezing.

Kaolinite-Silica Ratio Effect

The effect of the kaolinite-to-silica ratio on the pore evolution is shown in Fig. 3. The SEM images display the cross-section microstructures of the composites. For the pure silica sample (Fig. 3a), numerous small pores with an average diameter of 5 μm are formed. With the increase of kaolinite concentration (Figs. 3b-d), the pore size gradually increases to ~25 μm while the pore morphology remains similar. The connectivity of the pores is also improved with the kaolinite content increase.

The pore size evolution and pore morphology can be understood by ice nucleation and crystal growth during freezing. As discussed above, the porous microstructures are homogeneous throughout the composites due to the low temperature gradient in the samples and the high particle hindrance to the growing ice fronts. Added kaolinite content does not vary these two factors, thus the morphology of the samples is conserved. When frozen, water experiences a heterogeneous nucleation that is dependent on t he liquid-solid interfaces which lower the nucleation barrier. Apparently, silica nanoparticles are able to provide numerous nucleation sites that greatly promote the formation of numerous small pores. In contrast, the same volume kaolinite possesses less liquid-solid interfaces, thus a less number of larger pores are fabricated. Another reason for the formation of larger pores with kaolinite content is the particle size and shape. As shown in Fig. 3c insert, the walls of the pores consist of kaolinite platelets. Specifically, kaolinite particles form the framework of the sample with silica particles adsorbed to the faces. Due to their large sizes and the network structure, kaolinite particles are hardly repelled by the ice front during the freeze casting and subsequently lead to the increase of pore size and the conservation of pore morphology.

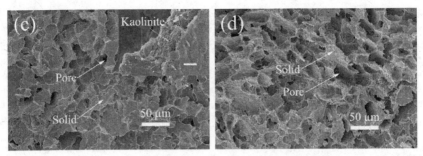

Figure 3. SEM images of the cross-sections of 2 K/min freeze cast samples: (a) 0% vol. kaolinite, (b) 6% vol. kaolinite, (c) 10% vol. kaolinite, (d) 14% vol. kaolinite. The insert image in Fig. 3c displays the kaolinite platelets in 10% vol. kaolinite sample.

The influence of kaolinite-silica ratio on the specific surface area can be explained by Fig. 4. The dashed line in Fig. 4a shows the specific surface area calculated by assuming the surface area is conserved from the kaolinite and silica particles upon freeze casting. Specifically, the surface area for the composite sample can be calculated using

$$S = \frac{S_{kaolinite}\, \rho_{kaolinite}\, \psi_{kaolinite} + S_{silica}\, \rho_{silica}\, \psi_{silica}}{\rho_{kaolinite}\, \psi_{kaolinite} + \rho_{silica}\, \psi_{silica}} \tag{2}$$

where S is the calculated specific surface area, $S_{kaolinite}$ and S_{silica} refer to specific surface area of kaolinite and silica nanoparticles, $\psi_{kaolinite}$ and ψ_{silica} refer to kaolinite volume fraction and silica volume fraction, $\rho_{kaolinite}$ and ρ_{silica} are density of kaolinite and silica, respectively. For calculation, the manufacturer-reported specific surface area of kaolinite (7 m^2/g) and calculated specific surface area of silica nanoparticles (115 m^2/g) were used respectively. The calculated curve agrees well with the experimental curve, indicating that the specific surface area is conserved when kaolinite particles and silica nanoparticles interact with each other. The loss of specific surface area in high kaolinite concentration samples is caused by the decrease of silica nanoparticle content. Fig. 4b shows the adsorption and desorption isotherm curve for 10% vol. kaolinite sample as a function of relative pressure. All the samples showed a similar adsorption and desorption isotherm behavior, thus only one sample is presented for the purpose of brevity. The hysteresis curve conforms to the classical type II isotherm, which confirms the porous microstructure of the fabricated sample.

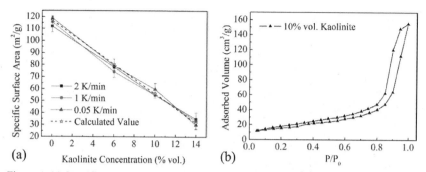

Figure 4. (a) Specific surface area for samples with different compositions and freeze cast at different freezing rates, (b) the adsorption and desorption curves for a 10% vol. kaolinite sample. The error bars indicate the range of the measured values.

Freezing Rate Effect

Three different freezing rates, 0.05 K/min, 1 K/min, 2 K/min, were applied to the freezing process in order to investigate their influence on the resulting composites. Since lower freezing rates leave more time for ice crystals to grow and allow particles to rearrange, pore size increase is assumed to occur with the decrease of the freezing rate. However, the microstructure evolution is not consistent with this hypothesis. Fig. 5 displays both cross-section and top surface of 10% vol. kaolinite composite frozen at the three freezing rates. The obvious pore size and pore morphology variation occurs on the top surface of the samples, while the difference between the cross-sections is not significant. Specifically, small pores on the top surface merge together to form larger ones when the freezing rate decreases. The connectivity of the top surface pores decreases dramatically and isolated narrow pores are finally formed in the 0.05 K/min sample. In contrast, pore size and pore morphology of the cross-sections of the three samples remain similar to each other. These trends in microstructure evolution can be understood by particle hindrance during the solidification. As mentioned above, both gelled network and kaolinite particle shape provide high hindrance force to effectively offset the solidified front growth. Thus, the interior pore size is determined by the distance between adjacent walls of the framework. Freezing rates have little impact on the interior microstructure compared to kaolinite-silica ratio effects. Particles on the top surface (free surface) are less constrained by the network and are relatively mobile. Therefore, the growing ice crystals on the free surface are less inhibited and the rearrangement of particles is more likely to occur, allowing the small ice crystals to grow into larger ones.

As shown in Fig. 4, freezing rates have very little effects on the specific surface area. This confirms that the specific surface areas of kaolinite platelets and silica nanoparticles are conserved during the freeze casting and are determined by the kaolinite-silica ratio.

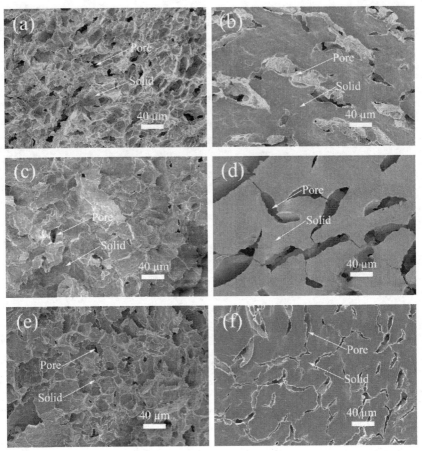

Figure 5. SEM images of the cross-sections and top surfaces of freeze cast 10% vol. samples. (a) and (b): 0.05 K/min, (c) and (d): 1 K/min, (e) and (f): 2 K/min.

Flexural Strength Test

The equibiaxial flexural strength test is applied to further understand the kaolinite-silica ratio and freezing rate effects on the resulting composites. The flexural strength is calculated using equation (1) and plotted in Fig. 6, where the Poisson's ratios for the kaolinite-silica porous composites (v_p) were obtained by the following equation[24]:

$$v_p = 0.5 - \left(1 - P^{2/3}\right)^{1.21} \Bigg/ 4\left[(1-s)\frac{(3-5P)(1-P)}{2(3-5P)(1-2v_0)+3P(1+v_0)} + s\frac{(1-P)}{3(1-v_0)}\right]$$

(3)

where $s = \dfrac{1}{1+e^{-100(P-0.4)}}$, P is the porosity, v_0 is Poisson's ratio for pore-free composite, which was obtained by linear interpolation based upon composition (Poisson's ratios for kaolinite and silica were assumed to be 0.17 and 0.45[25]).

As shown in the plot, the flexural strength of composites increases gradually to the maximum value as the kaolinite concentration grows to 14% vol., above which the strength decreases. Both pure silica and pure kaolinite samples have low flexural strength around 0.04 MPa, while kaolinite and silica particles work cooperatively to fabricate composites with reinforced strength. The strength improvement of the composites can be explained by the interaction and cooperation between kaolinite platelets and silica nanoparticles. For pure silica samples, the nanoparticles contact each other by point-to-point interaction. Evidence for this assumption is the conserved specific surface area after the freeze casting and the sphere-like nanoparticle shape. When a load is applied to the sample, stress cannot be transferred effectively throughout the sample due to the point contact. Thus, samples are crushed by concentrated stress. Added kaolinite particles are able to improve the contact between particles based upon their large size and plate-like shape, so that the concentrated stress can be rapidly distributed throughout the composite when a load is applied. In addition, the stiffness of kaolinite platelets itself is higher than that of packing of silica nanoparticles. Therefore, with the kaolinite concentration increase, the strength of the sample is gradually improved. The kaolinite-to-silica ratio that gives the highest strength is around the highest kaolinite concentration at which the gelled network can be formed. Above this value, the amount of silica nanoparticles is not enough to either induce the gel formation or bond kaolinite platelets together. The strength of particles decreases dramatically since there are not sufficient nanoparticles to bridge kaolinite platelets. Consequently, the strength of the composites decreases.

Decreasing the freezing rate slightly improves the flexural strength of the composites. This is because lower freezing rate leaves more time for particle rearrangement, although this process is limited by the gelled state and particle hindrance. Thus, the freezing rate effect on strength is less significant compared to kaolinite-silica ratio effect.

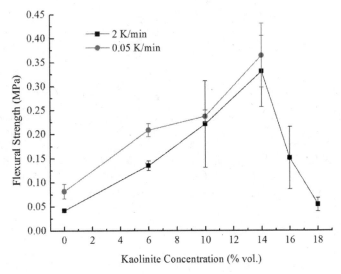

Figure 6. Equibiaxial flexural strength test for freeze cast composites with 0.05 K/min and 2 K/min freezing rates. The error bars indicate the standard deviation of the measured values.

CONCLUSIONS

This paper demonstrates the kaolinite-silica ratio and freezing rate effects on the microstructure evolution, specific surface area, and flexural strength of a freeze cast kaolinite-silica nanocomposite. Compared to freezing rate effect, solids loading has more significant influence on t he properties of composites due to the gelled state and plate-like kaolinite particles. With kaolinite concentration increase (solid concentration was kept at 18% vol.), pore size increases gradually while the pore morphology remains. Specific surface area depends on the kaolinite-silica ratio as the surface area of both kaolinite and silica particles are conserved during freeze casting. The flexural strength reaches the maximum when the composite consists of 14% vol. kaolinite and 4% vol. silica. The network of kaolinite platelets bonded by silica nanoparticles can effectively transfer the applied load, which eliminates the concentrated stress and improves the strength.

ACKNOWLEDGEMENT

The authors acknowledge financial support from the National Science Foundation, under grant no. CBET-0827246, and from the American Chemical Society Petroleum Research Fund, grant no. 47421-AC9. Assistance from the Nanoscale Characterization and Fabrication Laboratory of Virginia Tech is greatly acknowledged.

REFERENCE

[1] S. Deville, Freeze-Casting of Porous Ceramics: A Review of Current Achievements and Issues, *Adv Eng Mater,* **10**, 155-69 (2008).

[2] J. C. Han, L. Y. Hu, Y. M. Zhang, and Y. F. Zhou, Fabrication of Ceramics with Complex Porous Structures by the Impregnate-freeze-casting Process, *J Am Ceram Soc,* **92**, 2165-67 (2009).

[3] E. Munch, E. Saiz, A. P. Tomsia, and S. Deville, Architectural Control of Freeze-cast Ceramics Through Additives and Templating, *J Am Ceram Soc,* **92**, 1534-39 (2009).

[4] Y. M. Zhang, L. Y. Hu, and J. C. Han, Preparation of a Dense/Porous Bilayered Ceramic by Applying an Electric Field During Freeze Casting, *J Am Ceram Soc,* **92**, 1874-76 (2009).

[5] T. Waschkies, R. Oberacker, and M. J. Hoffmann, Control of Lamellae Spacing During Freeze Casting of Ceramics Using Double-side Cooling as a Novel Processing Route, *J Am Ceram Soc,* **92**, S79-S84 (2009).

[6] Y. H. Koh, J. H. Song, E. J. Lee, and H. E. Kim, Freezing Dilute Ceramic/Camphene Slurry for Ultra-high Porosity Ceramics with Completely Interconnected Pore Networks, *J Am Ceram Soc,* **89**, 3089-93 (2006).

[7] S. Deville, E. Maire, A. Lasalle, A. Bogner, C. Gauthier, J. Leloup, and C. Guizard, In Situ X-Ray Radiography and Tomography Observations of the Solidification of Aqueous Alumina Particle Suspensions. Part I: Initial Instants, *J Am Ceram Soc,* **92**, 2489-96 (2009).

[8] S. Deville, E. Maire, A. Lasalle, A. Bogner, C. Gauthier, J. Leloup, and C. Guizard, In Situ X-Ray Radiography and Tomography Observations of the Solidification of Aqueous Alumina Particles Suspensions. Part II: Steady State, *J Am Ceram Soc,* **92**, 2497-503 (2009).

[9] S. W. Sofie, Fabrication of Functionally Graded and Aligned Porosity in Thin Ceramic Substrates with the Novel Freeze-tape-casting Process, *J Am Ceram Soc,* **90**, 2024-31 (2007).

[10] S. W. Sofie and F. Dogan, Freeze Casting of Aqueous Alumina Slurries With Glycerol, *J Am Ceram Soc,* **84**, 1459-64 (2001).

[11] C. M. Pekor, P. Kisa, and I. Nettleship, Effect of Polyethylene Glycol on the Microstructure of Freeze-Cast Alumina, *J Am Ceram Soc,* **91**, 3185-90 (2008).

[12] K. Lu, C. S. Kessler, and R. M. Davis, Optimization of a Nanoparticle Suspension for Freeze Casting, *J Am Ceram Soc,* **89**, 2459-65 (2006).

[13] A. Szepes, A. Feher, P. Szabo-Revesz, and J. Ulrich, Influence of Freezing Temperature on Product Parameters of Solid Dosage Forms Prepared via the Freeze-casting Technique, *Chem Eng Technol,* **30**, 511-16 (2007).

[14] C. T. McKee and J. Y. Walz, Effects of Added Clay on t he Properties of Freeze-casted Composites of Silica Nanoparticles, *J Am Ceram Soc,* **92**, 916-21 (2009).

[15] L. L. Ren, Y. P. Zeng, and D. L. Jiang, Fabrication of Gradient Pore TiO2 Sheets by a Novel Freeze-tape-casting Process, *J Am Ceram Soc,* **90**, 3001-04 (2007).

[16] Y. Chino and D. C. Dunand, Directionally Freeze-cast Titanium Foam with Aligned, Elongated Pores, *Acta Mater,* **56**, 105-13 (2008).

[17] Y. Zhang, K. Zuo, and Y. P. Zeng, Effects of Gelatin Addition on the Microstructure of Freeze-cast Porous Hydroxyapatite Cerarmics, *Ceram Int,* **35**, 2151-54 (2009).

[18] M. C. Gutierrez, M. L. Ferrer, and F. del Monte, Ice-templated Materials: Sophisticated

Structures Exhibiting Enhanced Functionalities Obtained after Unidirectional Freezing and Ice-segregation-induced Self-assembly, *Chem Mater,* **20**, 634-48 (2008).

[19]J. C. Baird and J. Y. Walz, The Effects of Added Nanoparticles on Aqueous Kaolinite Suspensions I. Structural Effects, *J Colloid Interface Sci,* **297**,161-69 (2006).

[20]J. C. Baird and J. Y. Walz, The Effects of Added Nanoparticles on Aqueous Kaolinite Suspensions II. Rheological Effects, *J Colloid Interface Sci,* **306**, 411-20 (2007).

[21]M. Tourbin and C. Frances, A Survey of Complementary Methods for the Characterization of Dense Colloidal Silica, *Part Part Syst Char,* **24**, 411-23 (2007).

[22]M. Tourbin and C. Frances, Monitoring of the Aggregation Process of Dense Colloidal Silica Suspensions in a Stirred Tank by Acoustic Spectroscopy, *Powder Technol,* **190**, 25-30 (2009).

[23]ASTM Designation C1499–04 American Society for Testing and Materials International, West Conshocken, Pa (2004).

[24]M. Arnold, A. R. Boccaccini, and G. Ondracek, Prediction of the Poisson's Ratio of Porous Materials, *J Mater Sci,* **31**, 1643-46 (1996).

[25]Sebastian Lobo-Guerrero, The Elastic Moduli of Soils with Disperesed Oversize Particles, M.S. Thesis, University of Pittsburgh, PA, 2002.

CONTROLLING THE SIZE OF MAGNETIC NANOPARTICLES FOR DRUG APPLICATIONS

Luiz Fernando Cótica, Valdirlei Fernandes Freitas, Gustavo Sanguino Dias, Ivair Aparecido dos Santos
Department of Physics, Universidade Estadual de Maringá
Maringá, PR, Brazil

Sheila Caroline Vendrame, Najeh Maissar Khalil, Rubiana Mara Mainardes
Department of Pharmacy, Universidade Estadual do Centro-Oeste
Guarapuava, PR, Brazil

ABSTRACT
Water or alcohol dispersible magnetic nanoparticles allowed new opportunities for various biomedical applications as contrast agents for magnetic resonance imaging, magnetic field guided drug delivery, tumor treatment via hyperthermia and biomolecular separation and diagnostic imaging. Especially in drug delivery applications, drug transport through magnetic nanoparticles has been widely studied in an attempt to obtain particles with high carrier capacity of drugs, good stability in aqueous solutions, good biocompatibility with cells and tissues, among others. In this context, in this work, chemically synthesized magnetite nanoparticles (via the modified sol-gel processing), with different sizes, were obtained through heat treatments. Magnetic, structural (X-ray diffraction), microstructural (scanning electron microscopy) and cells toxicity properties of nanoparticles are carefully investigated.

INTRODUCTION
The synthesis of biocompatible magnetic nanoparticles has long been of interest in biomedical applications. The feasibility of many medical applications may strongly rely upon generating narrow size distribution and well dispersed nanoparticles in an aqueous or alcoholic solution.

With these characteristics, magnetic nanoparticles can be accurately manipulated by a magnetic field gradient. This "action at a distance", combined with intrinsic magnetic field penetration into human tissue, opens up many applications involving transport and/or immobilization of magnetic nanoparticles or magnetically added biological entities.[1] The main biomedical applications that use the magnetic characteristic of the nanoparticles include contrast agents for magnetic resonance imaging, magnetic field guided drug delivery, tumor treatment via hyperthermia and biomolecular separation and diagnostic imaging.[2-5]

Targeting these applications, a variety of magnetic nanoparticles composed of different atoms or ions with different magnetic moments have been synthesized. Inorganic nanoparticles such as Fe_3O_4 (magnetite), γ-Fe_2O_3 (maghemite) and spinel-structure magnetic particles (MFe_2O_4, M = Fe, Co, Ni, Mn) are increasingly finding applications in nanomedicine.[6] Most of the magnetic materials used in biomedical applications are 10–20 nm size range, i. e., they are in a single domain magnetic regime. This leads to the phenomenon of superparamagnetism. These materials respond to an applied external magnetic field but do not have any residual magnetization upon removal of the magnetic field.

Among the cited magnetic nanoparticles, the most used are magnetite nanoparticles. These nanoparticles offer many attractive possibilities in biomedicine. They have controllable sizes ranging from few nanometers up to tens of nanometers, which puts them in smaller dimensions when compared to some cells (10-100 μm), viruses (20-450 nm), protein (5-50 nm) or gene (2 nm wide and 10-100 nm long). This means that they near in size to biological entities of interest.[1,7,8]

Bulk and nanostructured magnetite crystallize with inverted spinel structure (space group *Fd-3m*) and the large oxygen ions are closely packed in a cubic arrangement, while smaller Fe ions fill the structural gaps (tetrahedral and octahedral sites). The tetrahedral and octahedral sites form two magnetic sublattices, A and B, respectively. The spins of sublattice A are antiparallel to those of B. The two crystal sites are very different and result in complex forms of exchange interactions of iron ions between and within the two types of sites. The structural formula of magnetite is $[Fe^{3+}][Fe^{3+},Fe^{2+}]_2O_4$ (AB_2O_4). This particular arrangement of cations in A and B sublattices characterizes the inverse spinel structure. With negative AB exchange interactions, the net magnetic moment of the magnetite is due to the B-site Fe^{2+} ions.[9]

In a biological statement, magnetite nanoparticles present low cytotoxicity and are well tolerated by the human body.[1,10] As a consequence, the drug release of magnetic drugs is able to concentrate these particles in a tumor if it is accessible through the arterial system and has a good blood supply. Therefore, the magnetic antitumor drug release occurs with fewer collateral effects and providing treatments shorter and less toxic.[6,11] However, for high efficiency in applications-specific biological function, magnetite must comply with certain criteria such as being spherical, biocompatible, superparamagnetic, have a narrow difference in size, high crystallinity, high magnetic saturation to provide a maximum control of the target and good dispersion in liquid.[5,7,12]

The composition, size and conditions for the synthesis of magnetite are determined by the target application, since this determines the physico-chemical and pharmacokinetic that the nanoparticles must be present.[6,13,14] Herein, we report a very simple route to synthesize uniform and size-controllable superparamagnetic magnetite nanoparticles with diameters smaller than 30 nm. Such magnetite nanoparticles are fabricated based on the thermal decomposition of a ferric nitrate/ethylene glycol solution. This attractive method is facile and convenient, requires neither harsh conditions nor extra modifiers, and provides an economical route to fabricate other functional inorganic nanomaterial. The structural a morphological properties of the nanoparticles were carefully studied via X-ray diffraction and scanning electron microscopy, respectively. The magnetic properties were obtained via vibrating sample magnetometry and the potential for biological applications was tested by in vitro citotoxic effects of magnetite nanoparticles essays.

EXPERIMENTAL

The magnetite nanoparticles were synthesized from a mixture of adequate amounts of ferric nitrate (Alfa Aesar) and ethylene glycol (Vetec). The solution was homogenized for one hour at room temperature and then subjected to heating until 90 °C. After heated, the solution was cooled to room temperature. To obtain nanoparticles with different sizes and distributions, the resulting material was heated in tunnel furnace at temperatures between 300 and 500 °C, under inert atmosphere (argon). X-ray diffraction (XRD) measurements were performed with a Shimadzu XRD 2000 diffractometer, using CuK_α radiation. The mean crystallite sizes were calculated from linewidth broadening relative to a metallic silicon standard by using Scherrer's formula. Morphological characterizations were made in a Shimadzu Superscan SS550 microscope. Magnetic characterizations were performed in a commercial vibrating sample magnetometer (VSM) Lakeshore at room temperature (300 K). The in vitro citotoxic effects of magnetite nanoparticles were studied using blood samples collected from three healthy volunteers. The blood was centrifuged and then plasma and the leukocytes layer were removed by aspiration. Erythrocytes were resuspended in a sodium phosphate buffer 50 mM, pH 7.4. Different concentrations of magnetite were incubated with an erythrocytes solution for 6 hours at 37 °C with constant mixing. Then, the solution was centrifuged and the content of hemoglobin released due to the cytotoxic action of magnetite nanoparticles in the supernatant was determined. This determination was made by measuring the absorbance at 540 nm in the absence or presence of magnetite (UV-Vis spectrometer Molecular Devices SpectraMax 190).

RESULTS AND DISCUSSION

Figure 1 shows the XRD patterns of the 300, 400 and 500 °C heat treated in argon atmosphere samples. All XRD patterns showed broadened X-ray lines, which were assigned to crystallized magnetite (JCPDS PDF19-0629) with inverted spinel structure (space group Fd-$3m$).[9] The mean crystallite sizes obtained for the Fe_3O_4 nanoparticles (using Scherrer's formula) were $t_{300} = 15$ nm, $t_{400} = 18$ nm and $t_{500} = 20$ nm for 300 °C, 400 °C and 500 °C heat treated samples, respectively.

Scanning electron microscopy images of magnetite nanoparticles heat treated at 300 and 500 °C are shown in Figure 2. The particles are in spherical shape with a 45 nm mean diameter and a narrow size distribution. Moreover, some magnetite nanoparticles partially superimposed onto the agglomerate core, i.e., the boundary limit of each nanoparticle is partially diffuse.

To gain information on the magnetic properties of the magnetite nanoparticles, the dependence of the room temperature magnetic field on magnetization (M x H curves) was investigated. Figure 3 shows M x H curves for the magnetite nanoparticles heat treated for 2 hours at 300 °C and 500 °C. These curves present typical superparamagnetic shapes. As already stated, nanoparticles usually are magnetic monodomains.[15-17] As can be seen, the saturation of magnetization (M_S) increases with the heat treatment temperature increasing. The increase in the M_S value leads to the magnetite nanoparticles has a magnetic saturation near the bulk magnetite (92 emu/g). The increase of M_S can be explained by considering that the heat treatment contributes to facilitate the exchange interaction between the magnetic sublattices of the magnetic core (Fe_3O_4) and to increase the magnetic core.[9]

In order to further explore the potential biological applications of the magnetite nanoparticles, we evaluated the cytotoxic effects of the 300 °C, 400 °C and 500 °C heat treated samples. Figure 4 shows a representative result obtained by spectroscopic measurements. The erythrocyte cytotoxicity assays showed no acute adverse effects comparing the obtained values for control samples and the magnetite containing samples. This characterizes the non-erythrocyte hemolysis in a wide concentration range up to 1.67 mg/mL after cell incubations. The negative result for hemolysis was confirmed by potassium ions quantification through ion-selective electrode, where there was no quantification of this ion. Thus, we can conclude that the magnetite nanoparticles have no cytotoxic activity on human erythrocytes.

In summary, the presented synthesis route allows control of the size and the magnetic properties of magnetite nanoparticles. The formation of the magnetite nanoparticles resulted in a multifunctional material with desired sizes, controlled magnetic properties and low toxicity properties. It is worth noting that these characteristics are very important in medical applications, such as DNA separation, magnetic drug targeting, immune detection, tissue engineering, hyperthermia in cancer treatment, detoxification of biological fluids, and magnetic guidance of particle systems for specific drug delivery processes.

CONCLUSIONS

In this paper were presented the results of detailed structural, microstructural, magnetic and biological studies in the magnetite nanoparticles obtained from a low cost thermal decomposition of a mixture of ferric nitrate and ethylene glycol. The samples were investigated by scanning electron microscopy, X-ray diffraction, magnetization measurements and in vitro citotoxic effects. Analysis of the crystallographic data revealed inverted spinel structure in all cases. Typical superparamagnetic behaviour was observed in magnetization measurements. The saturation of magnetization (M_S) increases with the heat treatment temperature increasing. The increase in the M_S value leads to the magnetite nanoparticles has a magnetic saturation near the bulk magnetite.

The erythrocyte cytotoxicity assays showed no acute adverse effects comparing the obtained values for control samples and the magnetite containing samples. The synthesis procedure used in this paper made it possible to tune the size and the magnetic properties of the magnetite nanoparticles. This tuning process is remarkably important in the design of materials for biological and biomedical applications, mainly drug delivery.

ACKNOWLEDGEMENTS

The authors would like to thank the CAPES (Proc. 082/2007), CNPq (proc. 307102/2007-2 and 302748/2008-3), and Fundação Araucária de Apoio ao Desenvolvimento Científico e Tecnológico do Paraná (Prot. 15727) Brazilian agencies for their financial support. S.C.V. also thanks the Fundação Araucária for the fellowship. We also gratefully acknowledge the instrumental research facilities provided by COMCAP/UEM.

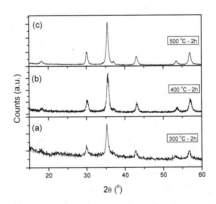

Figure 1 – X-ray diffraction patterns of the magnetite nanoparticles heat treated for 2 hours at (a) 300 °C, (b) 400 °C and (c) 500 °C. The patterns have been assigned to crystallites with spinel structure (*Fd-3m* space group) for all samples.

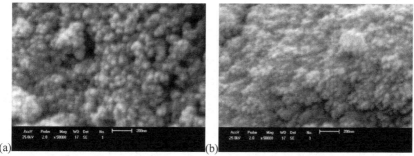

Figure 2 – Scanning electron microscopy images of the of the magnetite nanoparticles heat treated for 2 hours at (a) 300 °C and (b) 500 °C.

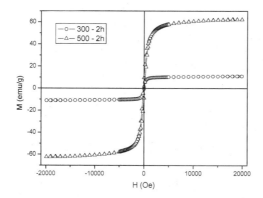

Figure 3 – Magnetization curves (M x H) of the magnetite nanoparticles heat treated for 2 hours at 300 °C and 500 °C.

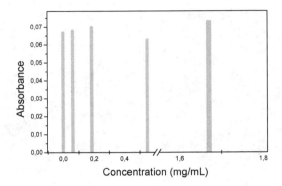

Concentration (mg/mL)

Figure 4 – Erythrocyte cytotoxicity assays for evaluates the cytotoxic effects of the 500 °C heat treated samples. These results are representative to all the other samples.

REFERENCES
[1]Q.A. Pankhurst, J. Connolly, S.K. Jones, J. Dobson, Applications of magnetic nanoparticles in biomedicine, *J. Phys. D.*, **36**, 167-181(2003).
[2]D.K. Kim, Y. Zhang, J. Kehr, T. Klason, B. Bjelke, M. Muhammed, Characterization and MRI study of surfactant-coated superparamagnetic nanoparticles administered into the rat brain, *J. Magn. Magn. Mater.*, **225**, 256–61 (2001).
[3]A.S. Lubbe, C. Bergemann, H. Riess, Clinical experiences with magnetic drug targeting: a phase I study with 40-epidoxorubicin in 14 patients with advanced solid tumors, *Cancer Res.*, **56**, 4686–93 (1996).
[4]D.C.F. Chan, D.B. Kirpotin, P. Bunn, Synthesis and evaluation of colloidal magnetic iron oxides for the site-specific radio frequency induced hyperthermia of cancer, *J. Magn. Magn. Mater.*, **122**, 374–8 (1993).
[5]A. Dyal, K. Loos, M. Noto, S.W. Chang, C. Spagnoli, K.V.P.M. Shafi, A. Ulman, M. Cowman, R.A. Gross, Activity of candida rugosalipase immobilizedon Fe_2O_3 magnetic nanoparticles, *J. Am. Chem. Soc.*, **125**, 1684–5 (2003).
[6]J. Yang, S.B. Park, H.G. Yoon, Y.M. Huh, S. Haam, Preparation of poly-caprolactone nanoparticles containing magnetite for magnetic drug carrier, *Int. J. Pharm.*, **324**, 185-190 (2006).
[7]R.H. Muller, S. Maaben, H. Weyhers, F. Specht, J.S. Lucks, Cytotoxicity of magnetite-loaded polylactide, polylactide/glycolide particles and solid lipid nanoparticles, *Int. J. Pharm.*, **138**, 85-94 (1996).
[8]Q.A. Pankhurst, J. Connolly, S.K. Jones, J. Dobson, Applications of magnetic nanoparticles in biomedicine, *J. Phys. D.*, **36**, 167-181 (2003).
[9]L.F. Cótica, I.A. Santos, E.M. Girotto, E.V. Ferri, A.A. Coelho, Surface spin disorder effects in magnetite and poly(thiophene)-coated magnetite nanoparticles, *J. Appl. Phys.*, **108**, 064325 (2010).
[10]O. Hafeli, M. Chastellain, *Nanoparticulates as drug carriers*, Imperial College Press (London), 397-411 (2006).

[11]R.B. Gupta, U.B. Kompella, *Nanoparticle technology for drug delivery*, Taylor & Francis Group (New York), **159**, 1-18 (2006).

[12]S. Guo, D. Li, L. Zhang, J. Li, E. Wang, Monodisperse mesoporous superparamagnetic single-crystal magnetite particles for drug delivery, *Biomaterials*, **30**, 1881-1889 (2009).

[13]T.Vo-Dinh, *Nanotechnology in biology and medicine: methods, devices and applications*, Taylor & Francis Gorup (Boca Raton), 762 (2007).

[14]D. Thassu, M. Deleers, Y. Pathak, *Nanoparticulate drug delivery systems*, Informa Healthcare (New York), 382 (2007).

[15]M. Jamet, V. Dupuis, P. Mélinon, G. Guiraud, A. Pérez, W. Wernsdorfer, A. Traverse, B. Baguenard, Structure and magnetism of well defined cobalt nanoparticles embedded in a niobium matrix, *Phys. Rev. B*, 62, 493-99 (2000).

[16]L. H. M. Fonseca, A.W. Rinaldi, A.F. Rubira, L.F. Cótica, S.N. de Medeiros, A. Paesano, Jr., I.A. Santos, E.M. Girotto, Structural, magnetic, and electrochemical properties of poly(*o*-anisidine)/maghemite thin films, *Mater. Chem. Phys.*, **97**, 252 (2006).

[17]R.A. Silva, M.J.L. Santos, A.W. Rinaldi, A.J.G. Zarbin, M.M. Oliveira, I.A. Santos, L.F. Cótica, A.A. Coellho, A.F. Rubira, E.M. Girotto, Low coercive field and conducting nanocomposite formed by Fe_3O_4 and poly(thiophene), *J. Solid State Chem.*, **180**, 3545 (2007).

CHEMICAL GROWTH AND OPTOELECTRONIC CHARACTERISTICS OF TiO₂ THIN FILM

[1]Chinedu Ekuma* ; [1]Israel Owate; [1]Eziaku Osarolube; [1]Evelyn Esabunor and [2]Innocent Otu

[1]Department of Physics, University of Port Harcourt, Pmb 51001, Rivers, Nigeria
[2]Department of Industrial Physics, Ebonyi State University, Abakaliki, Nigeria

ABSTRACT

The Chemical growth and characteristics of TiO₂ thin film doped with PbO have been carried out. Standard chemical bath thin film growth technique was applied in the deposition process. Different concentrations of PVP were used. Annealing was performed at temperatures of 150^0 C to 350^0 C. The Samples B, C and D of the same composition were annealed at different temperatures, and then, characterized using optical and electronic parameters. The result obtained showed that low annealing temperatures reduced absorbance, transmittance and reflectance. Also, the optical absorbance decreases with increasing wavelength within the UV-NIR regions of the electromagnetic spectrum. The Transmittance increases with increasing wavelengths whereas reflectance decreased with wavelengths. Data from density of states and dielectric constants indicated relative narrowing of the band gap of pure TiO₂ to ranges between 1.03-1.31 eV. This is due to PbO impurities. The properties indicate possibility of harnessing for applications such as in window materials and radiation collectors within visible region.

INTRODUCTION

The development of advanced materials for alternative and sustainable energy applications is an extremely active research area of great importance. In particular, much effort has been devoted to searching for new catalytic materials that can readily split water to generate hydrogen as an environmentally friendly fuel via photolysis using the abundant energy from sunlight [1]. Titanium dioxide (*TiO₂*) is one of the promising materials that suit these conditions. TiO₂ is not just a better photocatalysts in heterogeneous photocatalytic applications [2] because of its functionality, but also a promising material for photochemical applications [3]. The diverse advantages and promising applications offered by TiO₂ are numerous. They includes but not limited to: 1) photocatalysis of poisonous compounds [4]; 2) non – toxicity, long term stability and chemically inert [5]; 3) as a pigment in paints and the global oil crisis [6]; 4) photocatalysts for solar energy utilization [7 – 9].

The applications of TiO₂ with regards to its photoreaction efficiency are severely limited by its large intrinsic band gap (-3.0eV) that is capable of absorbing only in the ultraviolet region of the solar spectrum [6] with about 3% of the more abundant and important visible spectrum absorbed [2]. The effective utilization of visible light occupying the main part of the solar spectrum has been one of the important reasons for the increased study of optoelectronic properties of TiO₂. The most vital prerequisite for the enhancement of the solar energy conversion efficiency of TiO₂ is to harness it to absorb more in the visible region by methodically reducing the band gap below 2.0eV. Various doping materials have been used in an effort to improve the photoreactivity of TiO₂ and significantly shift its absorbing edge to the visible light region. Doping of different transition metal cations has been extensively studied [3,10,11]. Some groups have also tried doping of metal cations [12,13]; and, the substitution of a nonmetal atom such as nitrogen [14,15], sulfur [3] and fluorine [16] for oxygen has

---

* Present Address: Department of Physics and Astronomy, Louisiana State University, Baton Rouge, LA 70803, USA
Electronic Address: panaceamee@yahoo.com

also been investigated. However, to the best of our knowledge, there is little information on the deposition of PbO-doped TiO$_2$ thin films by chemical solution deposition (CSD).

In this study, we the successful deposition and characterization of transparent TiO$_2$ doped PbO thin films on soda-lime silica glass (SLSG) substrates by CSD on the pores of polyvinylpyrrolidone (PVP) matrix.

EXPERIMENTAL PROCEDURE

Preparation of Solutions: All the chemicals used were of analytical grade and all the solutions were prepared in doubly distilled water using standard methods.

Standard TiCl$_3$ solutions were prepared from its stock solution. 1M NaOH solution was prepared by dissolving 4g of its pellets in 100ml of distilled water. Gelatin (G-8) solution was prepared by dissolving 1.5gm of insoluble G-8 in 250ml of distilled water and stirred using magnetic stirrer to obtain a homogenous solution. Polyvinylpyrrolidone (PVP) solution was prepared by adding 4g of it to 400ml of distilled water. The mixture was initially insoluble but becomes soluble when stirred by a magnetic stirrer. Also, 1M Lead Acetate solution was prepared by dissolving 37.93g of lead acetate in 100ml of distilled water.

Experimentation: The soda-lime silica glass substrates were initially degreased in HCl for 24h, washed with mild soap, rinsed in water, cleaned with acetone and dried in open air. The deposition process was of two stages. 1) The initial deposition of TiO$_2$: The reaction bath contained 1M 5ml TiCl$_3$; 1M 3ml NaOH and 38ml gelatin carefully added in that order. The content of the bath was stirred to obtain a colorless homogenous mixture and substrate allowed in the reaction bath for 7h at 70°C. 2) The reaction bath contained 1.0M 10ml Pb(CH$_3$COO)$_2$.3H$_2$O; 40ml PVP and 6ml NH$_3$ added in that order. The mixture was carefully stirred to give colorless homogeneous mixture. Then, the substrates from first phase were carefully immersed into the reaction baths for 7h, and the temperature maintained at 70°C.

The slides were carefully removed from the bath, rinsed in distilled water and allowed to dry in open air. These samples were later annealed at different temperatures range of 150 to 350°C for 1h and then characterization for the optical and solid state properties. The normal incidence transmittance and absorbance spectrum of the deposited films were characterized using UNICO UV-2102 PC spectrophotometer in the UV-VIS-NIR region; and measurements were taken under the parametric conditions of normal incident, ambient temperature and uncoated microscopic glass slide was used as a blank. The Rutherford Back-scattering (RBS) characterization was also carried out to determine the stoichiometry of the film. The sample annealed at 150°C is labeled **B**, sample annealed at 250°C is labeled **C** and sample annealed at 350°C is labeled **D**.

RBS Measurement: The composition of the sample (unannealed) was determined by RBS using a Van de Graaff accelerator. A 2.20MeV, 3.00μC at 2.20nA $^4He^+$ ion beam at normal incidence was used, and backscattered particles were detected at 170°, with the number of atoms per unit surface area (areal density) determined by computer simulation. The method adopted in the analysis of the RBS is similar to that, which has been extensively described elsewhere [see 17,18].

RESULTS AND DISCUSSION

The results of the various optoelectronic properties of the developed films were as shown in Figures 1 to 4.

Optoelectronic Analysis: The spectral absorbance of the film as shown in Fig. 1 indicates that the absorption edge was shifted to the lower-energy region in the spectrum. The transmittance in the wavelength range 400 to > 900nm is observed to be relatively high (see Fig. 2) in the visible range and clear absorption edges of the films were observed. The high transmittances of the films may be attributed to the small particle size which reduces light scattering [2]. Critical analysis of these spectra, indicate that it is obvious that the films present relatively high optical quality, with an absorption in the visible region that is characterized by the typical interference pattern found when a transparent thin film is deposited onto a substrate of varying index of refraction.

From fig. 3, the films can be found to have direct optical band gap: 1.09eV for sample B, 1.03eV for sample C and 1.31eV for sample D; with band shifts of 0.06eV and 0.28eV in that order. Observe that the energy gaps were significantly affected by the varying annealing temperatures, as can be inferred from the band gap shifts. However, the most important inference is the relative narrowing of the band gap of pure titanium dioxide to well below 2eV where it becomes useful in the optical region of the electromagnetic spectrum. Also, since the photocatalysis of water occurs at band gap of around 1.23eV, the observed narrowing of band gap, no doubt, makes this material a good candidate for this very important application.

Compositional Analysis: The elemental composition and chemical states of the samples were analyzed by Rutherford backscattering (RBS) (see Fig. 4). It can be deduced that the stoichiometry of Oxygen (O): Titanium (Ti): Lead (Pb) is 0.758: 0.156: 0.086. The relatively low (trace) quantity of Pb in the film implies that Pb acts as an impurity element. The film deposited can thus be said to be $Ti_{0.156}O_{0.758}Pb_{0.086}$. The deposited film has thickness of 450nm which is far lower than ~520nm of pure TiO_2. This no doubt will enhance the photocatalytic and hence the biocidal activity of the film. The advantage of this is that by reducing the size of the TiO_2 particle, the surface area of TiO_2 increases leading to improvement of photoefficiency and thus, photocatalytic property because high surface area would make the surface of the particle more active to light and H_2O absorption. These observed trends of the properties of the film support the fact that they can be harnessed for possible application in collection of solar radiation.

CONCLUSION

We have successful deposited and characterized TiO_2 doped with PbO. Data from the density of states and the optoelectronic properties of the film reveal that there is significant band gap narrowing of the film towards the absorption edge of the visible light region; thereby making it very useful as a possible material for photocatalytic application. This can be harnessed for use as window materials and radiation collectors within the visible region. The annealing of the samples within the investigated temperature range has little or no effect on the overall intrinsic properties of interest of the doped material since the band gap is generally less than 2.0eV which makes all of them good window materials within this electromagnetic range of interest.

ACKNOWLEDGEMENT

The authors are grateful to all the Staff of Solid State Unit, Department of Physics and Astronomy, University of Nigeria, Nsukka; Engineering Material Development Institute (EMDI), Akure, Ondo State; and Centre for Energy Research and Development, Obafemi Awolowo University, Ile-Ife for their help in the cause of developing and characterization of the samples used in this work.

REFERENCES
1. Wenguang Z; Xiaofeng Q; Violeta I; Xing-Qiu C; Hui P; Wei W; Nada M. D; Tijana R; Harry M. M; Parans M.P; G. M. S; Hanno H. W; Baohua G; Gyula E and Zhenyu Z. (2009): Phys. Rev. Lett. 103, 226401.
2. Yong-Mu L; Ju-Hyun J; Jun-Hyung A; Young-Sun J; Kyung-Ok J; Kyu-Seog H and Byung-Hoon K. (2005): J. Cer. Proc. Res. 6(4), 302 – 304.

3. Umebayashi, T.; T. Yamaki and H. Itoh; K. A. (2002: Appl. Phys. Letts. 81(3).
4. Madhusudan, K.; Reddy, Sunkara V. Manorama, A. Ramachandra R. (2002): Mater. Chem. and Phys., 78, 239–245.
5. Ollis, D. F. and H. Al-Ekabi. (1993): *Photocatalysis Purification and Treatment of Water and Air,* Elsevier Science, New York.
6. Schmid, G., M. Baumle, M. Greekens, I. Heim, C. Osemann, T. Sawatowski. (1999): Chem. Soc. Rev. 28, 179–185.
7. Fujishima, A.; X. T. Zhang, and D. A. (2008): Tryk, Surf. Sci. Rep. 63, 515.
8. S. B. Zhang, (2002): J. Phys. Condens. Matter 14, R881.
9. Serpone, N., (2006): J. Phys. Chem. B 110, 24, 287.
10. Yamashita H., Y. Ichihashi; M. Takeuchi; S. Kishiguchi, and M. Anpo. (1999): J. Synchrotron Radiat. **6**, 451.
11. Karakitsou, K. E. and X. E. Verykios, (1993): J. Phys. Chem. **97**, 1184.
12. Li, G.H., L. Yang, Y.X. Jin, and J.D. Zhzng, (2000): Thin Solid Films 368, 163-167.
13. Gracia, F., J.P. Holgado, F. Yubero, and A.R. González- Elipe, (2002): Surf. & Coat. Tech. 158-159, 552-557.
14. Asahi, R.; T. Morikawa; T. Ohwaki; A. Aoki, and Y. Taga: (2001): Science **293**, 269.
15. Morikawa, T.; R. Asahi, T. Ohwaki, A. Aoki, and Y. Taga. (2001): Jpn. J. Appl. Phys., Part 2 **40**, L561.
16. Hattori, A.; M. Yamamoto, H. Tada, and S. Ito. (1998): Chem. Lett., 707.
17. Simone, B; Eros A; Marco B and G L. (2002): Phys. Rev. B, 66, 045202.
18. Ferroni M.; V. Guidi; G. Martinelli; E. Comini; G. Sberveglieri; D. Boscarino and G. Della Mea (2002): J. Appl. Phys. 88, 2.

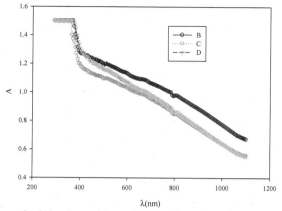

Fig. 1: The graph of Absorbance (A) as a function of wavelength (λ) for TiO$_2$-PbO thin film

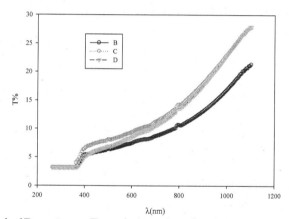

Fig. 2: The graph of Transmittance (T) as a function of wavelength (λ) for TiO$_2$-PbO thin film

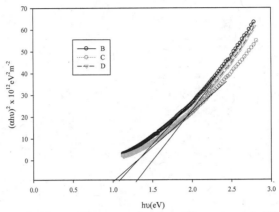

Fig. 3: The graph of Density of states as a function of Photon energy for TiO₂-PbO thin film

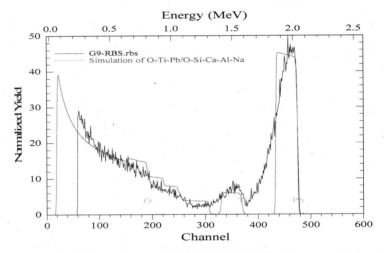

Fig. 4: The Rutherford Backscattering (RBS) of TiO₂-PbO thin film of the unanealed sample.

SYNTHESIS OF MANGANESE OXIDES NANOCOMPOUNDS FOR ELECTRODES IN ELECTROCHEMICAL CAPACITORS

R. Lucio[1], I. Gómez[1], L. Torres[1]
Affiliation 1, Materials Science Laboratory, Chemistry S. Faculty, UANL, San Nicolás de los Garza, N.L., México

P. Elizondo[2]
Affiliation 2, Synthesis Laboratory, Chemistry S. Faculty, UANL, San Nicolás de los Garza, N.L. México

ABSTRACT

Nanocrystals of Mn_3O_4 with different shapes and diameters under 100 nm were prepared using normal micelles as templates. Modifying the solution conditions allowed us to control the particle shape, inducing changes in micelles shape as the packing parameter predicts. The key factors to control the nanostructure of particles are the interactions between micelles and Mn^{+2} ions and the molar ratio of SDS/Mn. When the molar ratio of SDS/Mn was relatively high, nanoparticles acquire the micelles shape, and when the molar ratio was relatively low, a transition from spherical to cylindrical micelles is predicted by the packing parameter and results in rod-shaped nanocrystals. All the samples obtained shown pseudo capacitive behavior.

INTRODUCTION

Nowadays, nanomaterials chemistry has focused in the synthesis of nanocrystals with controllable shapes, due to their size and shape-dependent properties. Thus, nanocrystals are designed for many potential applications like optoelectronics, catalysis, energy storage, chemical sensing, among others [1,2]. Among the synthetic routes to prepare nanomaterials, inverse micelles have been widely used as nanoreactors to control size and shape of nanoparticles by modulating the micelle shape through modifying the solution conditions. The weakness of this method is that the organic solvents used and products are prepared commonly in diluted form, which may difficult large scale production [2,3]. On the other hand, surfactant-water systems should form normal micelles with several shapes in a wide range of conditions.

Although we can predict the micelles shape with a phase diagram of surfactant-water system, one can achieve control of micelles shape by the well known packing parameter that depends on the molecular structure of surfactants and predicts the micelles shape at equilibrium conditions. It is defined by $P = Vc/alc$, where Vc and lc are the volume and the length of surfactant tail and a is the area per headgroup. Thus, for a different value of P there is a specific shape for the micelles; $0<P<1/3$ is for spherical, $1/3<P<1/2$ is for cylindrical and $1/2<P<1$ is for a bilayer [4]. For a given surfactant-water system Vc/lc is constant and a depends on the electrostatic repulsions between headgroups at the micelle surface, can be modified by the solutions conditions. For example, adding an electrolyte to a system with an ionic surfactant will decrease due to a decrease in electrostatic repulsions of headgroups which results in an increase of P. Thus, a change of the micelles shape is induced [5,6].

Attempts to control the shape of nanostructured silica-based materials through the packing parameter were done by Stucky et al [7]. The authors chose surfactants with different molecular structure to vary P, in order to control the shape of nanoparticles. However, until now, there is no study pointing out how the morphology of the micelles can be modulated to achieve nanomaterials with different shapes, by modifying the solution conditions.

Hausmannite (Mn_3O_4) is an important material due to its wide range of applications such as energy storage, high-density magnetic storage media, oxygen storage, catalysts, ion exchange, molecular adsorption, varistors and solar energy transformation [8-10]. The above have motivated the development of new synthetic routes to prepare Mn_3O_4 in nanometer scale.

Here, we report the synthesis of nanocrystalline shapes of Mn_3O_4 using normal micelles. Modifying the solution conditions like the molar ratio surfactant/precursor allowed us to control the micelles shape through modulating the repulsion forces between headgroups, inducing changes on micelles shape as the packing parameter predicts, producing particles with different shapes that replicates micelles shapes. Later we present a study about capacitive properties of the samples prepared in order to obtain electrochemical supercapacitors.

EXPERIMENTAL

Synthesis of Mn_3O_4

An appropriate amount of sodium dodecylsulfate (SDS, Alfa-Aesar 99%) was dissolved in deionized water, then $MnCl_2 \cdot 4H_2O$ (Alfa-Aesar, 99%) was added. The solution was heated at 70°C and an appropriate amount of NaOH (J.T. Baker) was added. The reaction time was 24 hrs. The products were filtrated and washed several times with deionized water and dried at room temperature. Table 1 shows the different compositions studied.

Table 1 Compositions of prepared experiments

Experiment	SDS (%w)	*SDS/Mn	*Mn/OH
S1	20	2.0	1.2
S2	5	2.0	1.2
P1	10	0.9	1.2
P2	5	0.5	1.2

*Molar ratio

Characterization

TEM images were obtained in a JEOL 2010 instrument, operating at 200kV. The sample was dispersed in acetone, deposited in a copper grid and dried at room temperature. FTIR analyses were made in a Perkin Elmer Paragon 1000 PC. Pellets were prepared with KBr in a ratio of 1:100 (sample:KBr). Spectra were collected with a resolution of 4cm^{-1} and 500 scans. XRD analysis were done in a Siemens D5000 CuKα (λ=1.5418Å). Cyclic Voltammetry were carried out in a Gamry PC4/759 equipment. Electrodes as tablet with Samples synthesized, conductive carbon (Degussa) and politetrafluoroethilene (PTFE Aldrich), at 70-25-5 %w respectively were prepared, with thickness already of 80-100 µm. An Ag/AgCl as reference electrode and platinum wire as contra electrode were used. Voltammograms were recorded with a solution prepared from Na_2SO_4 0.1M as supporting electrolyte.

RESULTS AND DICUSSION

Figure 1a shows XRD patterns (JCPDS 24-0734) of these nanocrystals, corresponding to Mn_3O_4. The main reflections can be indexed to centered tetragonal structure with $I4_1/amd$ space group. The Mn^{+2} and Mn^{+3} ions occupy the tetrahedral and octahedral sites respectively. However, a small diffraction peak corresponding to MnOOH is present (JCPDS 41-1379). Figure 1b shows the FTIR spectra of these compounds. The band at 615 cm^{-1} corresponds to Mn-O stretching modes in tetrahedral sites. The band at 500 cm^{-1} is characteristic of distortion vibration of Mn-O in an octahedral environment and the one at 415 cm^{-1} corresponds to vibration of Mn^{+3} in octahedral sites [11]. These bands are characteristic of Mn_3O_4 phase in agreement with XRD results. Bands located at 3420 and 1624 cm^{-1} correspond to adsorbed water and the bands located at 1224, 1152, 1110, 1086 cm^{-1} correspond to -OSO_3 group [11,12]. Finally, the bands at 2919 cm^{-1} and 2849 cm^{-1} are due to the C-H stretching modes, asymmetrical and symmetrical respectively and correspond to surfactant chains [13] meaning that surfactant is present in the prepared products.

The chemical reactions under these conditions can be described as follows:

$$Mn^{+2} + 2OH^- \rightarrow Mn(OH)_2 \qquad (1)$$
$$3Mn(OH)_2 + 0.5O_2 - Mn_3O_4 + 3H_2O \qquad (2)$$

First, the manganese hydroxide is formed, and then the oxygen dissolved in the solution promotes the formation of Hausmannite. The good crystallinity of the nanoparticles it is attributed to a small size of $Mn(OH)_2$ that allows an efficient oxidation by O_2 [14].

The experimental conditions of S1 and S2 are presented in Table 1. For both experiments the surfactant is in excess with respect to $MnCl_2$, but differs in the surfactant concentration (20 and 5 w% respectively). At concentration of 15%w and 5%w at 70°C cylindrical and spherical micelles are predicted for SDS-water system [15] respectively. Figure 2a and 2b show that for experiment S1 and S2 the particles are rods and spheres respectively, both with diameter under 100nm. This means that products are replicates of micelles morphology under these conditions, since a composition of 20%w cylindrical micelles are expected and for a composition of 5 %w spherical micelles are predicted. Thus, micellar templates contribute to the nanostructure of products. SDS is an anionic surfactant, so SDS micelles are negatively charged and can interact with Mn^{+2} ions by electrostatic forces. These interactions promotes that manganese ions stay around micelles and when OH$^-$ ions were added to solution, the nanostructure of $Mn(OH)_2$ is a replicate of the micelles morphology. Finally, manganese hydroxide is oxidized by O_2 to Mn_3O_4.

When $MnCl_2$ is in excess with respect to SDS (P1), nanorods of Mn_3O_4 are obtained (figure 2c) with a diameter under 100nm, despite that a composition of 10%w spherical micelles are also predicted for SDS-water system [15]. This result can be explained by the fact that electrostatic repulsion forces between headgroups (at the micelles interface) are decreased due to an excess of Mn^{+2} ions present in the solution, thus a decrease and as packing parameter predicts, a transition from spherical to cylindrical micelles is expected [5,6]. Therefore, as Mn^{+2} ions surround cylindrical micelles and OH$^-$ ions added to the solution, the nanostructure of $Mn(OH)_2$ is a replicate of the micelles morphology. In experiment P2 there is an excess of Mn^{+2} ions with respect to SDS, a transition from spherical to cylindrical micelles is also expected and as consequence rod-shaped nanocrystals under 100 nm were obtained after $Mn(OH)_2$ is formed and oxidized to Mn_3O_4 (figure 2d).

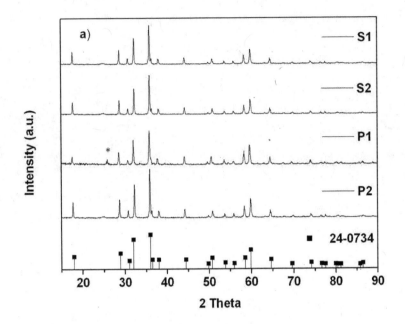

Figure 1a. Powder XRD patterns of the prepared samples that corresponds to Mn$_3$O$_4$ (JCPDS 24-0734). The peak labeled with "*" correspond to MnOOH (JCPDS 41-1379)

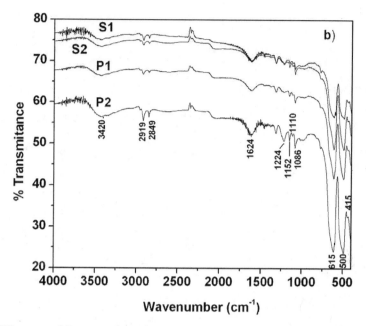

Figure 1b. FTIR spectra of the prepared samples.

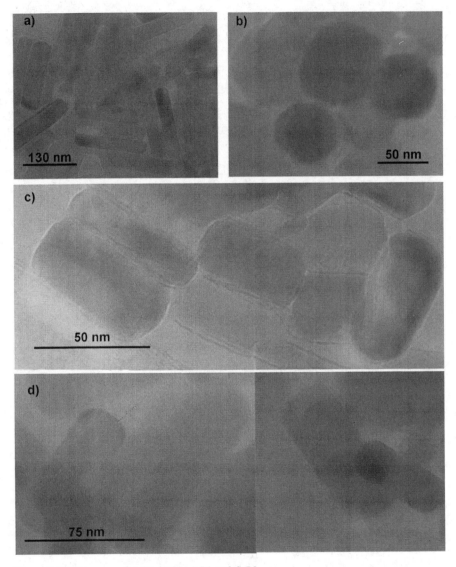

Figure 2. TEM images of a) S1, b) S2, c) P1 and d) P2.

The molar ratio of SDS/Mn is a key factor to control the nanostructure of Mn_3O_4. While the molar ratio of SDS/Mn equal to 2, the nanostructure of the particles are replicates of the micelles morphology. Apparently there is no transition from spherical to cylindrical micelles due to Mn^{+2} ions concentration insufficient to decrease the repulsion forces between headgroups. When there is an excess of Mn^{+2} ions with respect to SDS, a transition from spherical to cylindrical micelles may occur, as the packing parameter predicts, leading to rod-shaped nanocrystals. These results are in agreement with previous observations [6] for a molar ratio under 1 (surfactant/salt) where a transition from spherical to cylindrical micelles is induced.

Interactions between surfactant and precursors are essential to control the nanostructure of particles. For example, Wang et al. [16] prepared Zn, Sn and Ni oxides; they used CTAB a cationic surfactant and inorganic salts as precursors. They showed particle agglomerates of the metal oxides but no sphere or rod-shaped nanocrystals were observed. We considered that repulsions forces between CTAB and Zn^{+2}, Sn^{+4} and Ni^{+2} ions do not allow an appropriate interaction between micelles and metal ions. In our work, the interaction between negatively charged micelles and Mn^{+2} ions permits that micelles surface acts as nucleation sites and influence particle growth. Therefore, micelles could act as templates for the syntheses of nanostructured materials.

Figure 3a-d shown voltammograms for prepared samples, S1, S2, P1 and P2 respectively. Potential window used was from -0.1 to 0.8 V (E vs Ag/AgCl).

Samples S1, P1 and P2, which correspond to cylindrical morphology, present pseudo capacitive behavior at scan rates of 2, 5 and 10 mV/s (figures 3a, 3c and 3d respectively). These effects are observed due rectangular and symmetric form of the curves respect to baseline [17-20]. At higher scanning rates of analyses this behavior is loss.

The sample S2, corresponding to spheroid morphology, presents pseudo capacitive behavior at scan rates of 2 and 5 mV/s, as can be observed in the rectangular and symmetric form of the curves respect to baseline [17-20].

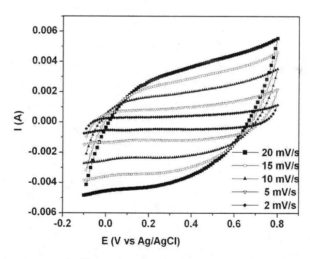

Figure 3 a. Voltammogram for S1 sample, using Na$_2$SO$_4$, 0.1M as electrolyte.

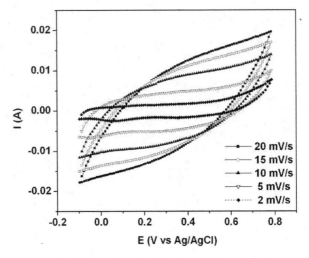

Figure 3 b. Voltammogram for S2 sample, using Na$_2$SO$_4$, 0.1M as electrolyte.

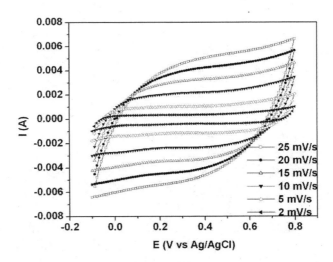

Figure 3 c. Voltammogram for P1 sample, using Na$_2$SO$_4$, 0.1M as electrolyte.

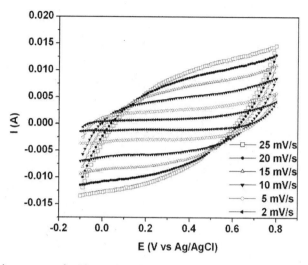

Figure 3 d. Voltammogram for P2 sample, using Na_2SO_4, 0.1M as electrolyte.

Values of specific capacitance for samples prepared are shown in Table 2. Specific capacitance calculated values due double layer electric are presented too. These were obtained by multiplication specific area of the sample by average value of double layer electric capacitance ($0.2 \ F/m^2$) [21].

Table 2 Specific capacitance values at different rates of analyses.

Sample	C (F/g) 2 mV/s	C (F/g) 5 mV/s	C (F/g) 10 mV/s	*Double Layer (F/g)	P (F/g) 2 mV/s	P (F/g) 5 mV/s	**V. P. (V)
S1	38	33	30	5.4	32.6	27.6	0.9
S2	40	35	-	4.8	71.2	56.2	0.7
P1	37	35	32	4.4	32.6	30.6	0.9
P2	39	33	28	3.4	35.6	29.6	0.9

C: Specific capacitance. P: specific pseudo capacitance. *specific capacitance due double layer. **Potential Window.

The capacitances due to double electric layer for all samples are showed in Table 2. These values are lower as experimental measures, and this effect allows us to conclude that specific capacitance of these materials comes from pseudo capacitive mechanism [22-25].

CONCLUSIONS

This work describes a simple method to control the nanostructure of Mn_3O_4, using micelles as templates. The interactions between micelles and Mn^{+2} ions and the molar ratio of SDS/Mn are the key factors to control nanoparticle shape. At a relative high molar ratio of SDS/Mn particles acquire the same morphology of micelles. When spherical micelles were present in the system and a low molar ratio of SDS/Mn was used rod-shaped nanocrystals were obtained, meaning that through modulating the concentration of Mn^{+2} ions, micelles shape changes, as the packing parameter predicts, influencing the particle shape. The samples prepared present pseudo capacitive behavior at scan rates lower of 15 mV/s.

Acknowledgments

Authors express their gratitude to Chemistry Science Faculty, PAICYT-UANL and CONACYT for economical support for the development of this work.

References

[1] Hao E., Bailey C., Schatz G.C. , Hupp J. T., Li S. Synthesis and Optical Properties of Branched Gold Nanocrystals, *Nano Lett.* Vol 4, 2004, p 327-330.

[2] Gao L., Wang E., Lian S. , Kang S., Lan Y. , Wu D. Microemulsion-directed synthesis of different CuS nanocrystals, *Solid State Commun.* Vol. 130, 2004, p 309-312.

[3] Lisiecki I., Pileni M.P. Spin crossover compound in nano-confined reversed micelle system, *J. Am. Chem. Soc.* Vol. 115, 1993 p 3887-3896.

[4] Nagarajan R. Molecular Packing Parameter and Surfactant Self-Assembly: Neglected Role of the Surfactant Tail, *Langmuir.* Vol. 18, 2002, p 31-38.

[5] Yan Y., Xiong W., Li X., Lu T., Huang J., Li Z., Fu H. Molecular Packing Parameter in Bolaamphiphile Solutions: Adjustment of Aggregate Morphology by Modifying the Solution Conditions, *J. of Phys. Chem. B.* Vol. 111, 2007, p 2225-2230.

[6] Hassan P.A., Raghavan S.R., Kaler E.W. Microstructural Changes in SDS Micelles Induced by Hydrotropic Salt, *Langmuir.* Vol. 18, 2002, p 2543-2548.

[7] Huo Q., Margolese D. I., Stucky G. D. Surfactant control of phases in the synthesis of mesoporus with periodic 50 to 300Å. *Chem. Mater.* Vol. 8, 1996, p 1147-1160.

[8] Dubal D.P., Dhawale D.S, Salunkhe R.R., Pawar S.M., Fulari V.J., Lokhande C.D. A novel chemical synthesis of interlocked cubes of hausmannite Mn_3O_4 thin films for supercapacitor application, *J. Alloys and Compounds.* Vol. 484, 2009, p 218-221.

[9] Dai Y., Wang K., Xie. Magnetic fluctuation and frustration in new iron-based layered $SrFe_{1-x}Co_xAsF$ superconductors, *J. Appl. Physics* Lett. Vol. 90, 2007, p 104102-3.

[10] Apte S.K., Naik S.D., Sonawane R.S., Kale B.B, Pavaskar N., Mandale A.B., Das B.K.. Nanosize Mn_3O_4 (Hausmannite) by microwave irradiation method, *Mater. Res. Bull.* Vol. 41, 2006, p 647-654.

[11] Gibot P., Laffont L. Hydrophilic and hydrophobic nano-sized Mn3O4 particles, *J. Solid State Chem.* Vol. 180, 2007 p 695-701.

[12] Luca V., Watson J. N., Ruschena M., Knott R. B. Anionic Surfactant Templated Titanium Oxide Mesophase: Synthesis, Characterization and Mechanism of Formation, *Chem. Mater.* Vol. 18, 2006, p 1156-1168.

[13] Snyder R.G , Strauss H.L. Carbon-hydrogen stretching modes and the structure of n-alkyl chains. 1. Long. Disordered chains, *J. of Phys. Chem.* Vol. 86, 1982, p 5145-5150.

[14] Vázquez-Olmos A., Redón R., Mata-Zamora M.E., Morales-Leal F., Fernández-Osorio A.L., Saniger. Structural and magnetic study of Mn_3O_4 nanoparticles, *J. M. Rev. Adv. Mater. Sci.* Vol. 10, 2005, p 362-366.

[15] Reiss-Husson F., Luzzati V. Polymorphism of Lipid-Water Systems: Epitaxial Relationships, Area-per-Volume Ratios, Polar-Apolar Partition, *J. Phys. Chem.* Vol. 68, 1964, p 3504-3511.

[16] Wang Y., Ma C., Sun X., Li H. Preparation of nanocrystalline metal oxide powders with the surfactant mediated method, *Inorg. Chem. Commun.* Vol. 5, 2002, p 751-755.

[17] Machefaux E., Brousse T., Bélanger D., Guyomard D., Supercapacitor Behavior of New Substituted Manganese Dioxides, *Journal of Power sources*, Vol 165, 2007, p 651-655.

[18] Conway B. E., Electrochemical Supercapacitors. *Scientific Fundamentals and Technological Applications.* Kluwer Academic/ Plenum Publishers. New York. 1999.

[19] Simon P., Gogotsi Y., Materials for Electrochemical Capacitors, *Nature Materials*, Vol 7, 2008, p 845-854.

[20] Jayalaskshmi M., Balasubramanian K., Simple Capacitors to Supercapacitors- An Overview. *International Journal of Electrochemical Science.* Vol 3, 2008, p 1196-1217.

[21] M. Winter, R. J. Brod., What Are Batteries, Fuel Cells, and Supercapacitors?, *Chem. Rev.* Vol 104, 2004, p 4245-4269.

[22] Toupin M., Brousse T., Bélanger D., Influence of Microstructure on the Charge Storage Properties of Chemically Synthesized Manganese Dioxide. *Chemical Materials.* Vol 14, 2002, p 3946-3952.

[23] Subramanian V., Zhu H., Vajtai R., Ajayan P. M., Wei B. Hydrothermal Synthesis and

Pseudocapacitance Properties of MnO$_2$ Nanostructures. *Journal of Physical Chemistry B*. Vol 109, 2005, p 20207-20214.

[24] Toupin M., Brousse T., Bélanger D. Charge Storage, Mechanism of MnO$_2$ Electrode Used in Aqueous Electrochemical Capacitor. *Chem. Mater.* Vol 16, 2004, p 3184-3190.

[25] Jiang J., Kucernak A., Electrochemical Supercapacitor Material Based on Manganese Oxide: Preparation and Characterization. *Electrochimica Acta.*, Vol 47, 2002, p 2381-2386.

Nanotechnology for Energy, Healthcare and Industry

FINITE ELEMENT MODELING OF SAPPHIRE PHOTONIC CRYSTAL FIBERS

Neal T. Pfeiffenberger and Gary R. Pickrell
Virginia Tech
Blacksburg, VA, USA

ABSTRACT
    This paper presents the finite element modeling of a unique sapphire photonic crystal fiber structure. The structure consists of six holes symmetrically arranged within the outer single crystal sapphire layer surrounding the solid single crystal sapphire core. The single crystal sapphire fibers produced were approximately 200μm in diameter with hole sizes in the range of 10μm. This represents the first time that a photonic crystal fiber has been fabricated and modeled in single crystal sapphire. The modeling work focuses on the optimization and modal analysis of this fiber using Comsol Multiphysics 4.0a.

INTRODUCTION
    Today's optical fibers are geared towards both communications and sensing industries. Most communication fibers are composed of a glass core with a higher refractive index than that of the solid glass cladding. This minimizes transmission losses by confining the light to the central core. One of the major goals in the data sensing market is the need for new high-temperature fiber designs. These types of designs are useful in a wide variety of applications including defense, civil infrastructure monitoring, and down-hole oil well sensing. Design of the appropriate fiber structure is critical in today's sensor and data transmission markets. Among the many advantages, optical fibers are immune to electromagnetic interference. This helps to reduce the noise during signal transfer, which gives a significant improvement over conventional electrical cabling. This is an important consideration in many high temperature-sensing applications.
    The fibers under study in this work are referred to as sapphire photonic crystal fibers (PCF's). They are composed of sapphire rods, which make up the core surrounded by air. Sapphire photonic crystal fibers belong to a branch of micro-structured optical fibers (MOFs)[1,2,3]. These MOF fibers are divided into two main categories. The first group operates via total internal reflection (TIR). Light is guided from the strong contrast between the core and cladding index (1.74618 for α-Al_2O_3 vs. 1.0 for air at a free-space wavelength of 1.55μm.) The second group operates under a photonic band gap produced by a periodic cladding region. The fiber in this work is guided by TIR and falls into the first group of photonic crystal fibers.
    Micro-structured optical fibers have been characterized and fabricated since the early work of Knight, Birks, Russell and Atkin[1,2,3,4,5]. The photonic crystal, which belongs to this class of micro-structured optical fibers, generally consists of a pure silica core surrounded by an ordered array of air holes in the cladding region. This regular or periodic pattern gives rise to the term crystal, which continues down the length of the fiber. Also under the umbrella of micro-structured optical fibers is a new type of holy fiber, called the random hole optical fiber (RHOF)[6], which has thousands of longitudinal holes that surround a solid central optical core. These holes vary in size, spatial location, and number throughout the optical axis[7]. These fibers are interesting for both sensing applications[8,9,10,11,12] as well as the fiber to the home application in the communications sector[13].

Photonic crystal fibers have many properties that are generating a significant amount of interest. Some of these properties include an extremely small core size, single mode guidance over a large wavelength range, and even guidance through a hollow core[4,5,14,15]. As previously stated, if a periodic lattice that has the correct size and spacing of the holes is formed then photonic band gap similar to that found in a semiconductor can be produced. This optical photonic band gap operates on the principle that periodic arrangement of a material with a given refractive index difference between that of the background material will not allow light of a certain range of wavelengths to be transmitted. If one of the holes in the lattice is removed, then this will confine a range of wavelengths from operating anywhere besides in this "defect". A photonic band gap will not, however, occur if the size and/or spacing of the holes are not in correct relation to the wavelength. These "non" photonic band gap photonic crystal fibers can still strongly confining light by lowering the average effective index of the cladding region. There have been many useful fibers that have come from these types of fibers but all of them are based on silica. Silica fibers are not useful at high temperatures due to fiber crystallization, fiber reactions with the environment, and glass creep under stress. Many sensing applications require long wavelength detection capabilities greater than the 1.6μm range that silica offers.

Sapphire offers mechanical characteristics that are superior to silica. Single crystal sapphire has a high melting point (2054 °C), high laser damage threshold[16], excellent chemical resistance in corrosive environments[17] and high hardness. Single crystal sapphire fibers are generally fabricated by the laser heated pedestal growth (LHPG) method[18]. Optical properties for single crystal sapphire include a transmission window that extends beyond 5μm (with about a 20% transmission at wavelengths as long as 6μm), good nonlinearity in the $n_2 \approx 3 \times 10^{-20}$ m^2/W range[19] and high availability. Single crystal sapphire fibers are already being researched and commercially used in the sensing fields as both intrinsic and extrinsic interferometers, polarimetric devices, and birefringence-balanced lead-insensitive sensors. These have all been demonstrated for measurement of physical parameters at high temperatures[20,21,22,23,24].

Improvements in the technology are still required if the high temperature capabilities of single crystal sapphire fibers are to be fully exploited. This is due to the lack of high temperature claddings on the sapphire optical fibers. The lack of cladding on the single crystal sapphire fibers results in higher environmental vulnerability of the single crystal sapphire fibers due to fluctuations in the optical signal as a result of materials adsorbed onto the fiber surface. This creates reactions at the sapphire surface, increasing the modal volume, and resulting in higher phase difference between low and high order modes. The loss from higher order modes will also create increased sensitivity due to fiber bending.

This paper is focused on studying the structure and evanescent waves of the first cladded sapphire photonic crystal fiber through the use of Finite Element Modeling (FEM). FEM modeling is so useful for this early development work because it accuracy helps to predict the optimal positioning and design of the fiber in a very efficient manner. This work is an extension of work published previously by our group[25] which adds the computer modeling analysis.

EXPERIMENTAL SECTION

The fibers in the following sections are based off both a single crystal sapphire rod and a single crystal sapphire rod surrounded by a ring of 6 other single crystal sapphire rods. The process for creating the 6-ring fiber is as follows. Single crystal sapphire rods, with an outer diameter of 70μm, were cleaved into 7 pieces, approximately 15cm each. Next, the fibers were cleaned in a series of steps which included rinsing in alcohol, distilled water, acetone, and diluted phosphoric acid. After the cleaning, the fibers were assembled in a pattern that consisted of a central fiber surrounded by six outer fibers. This process forms a symmetric ring around the central fiber as seen in the schematic of the fiber placement shown in Figure 1[25].

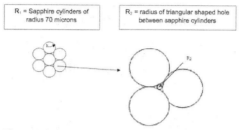

Figure 1. Schematic of fiber used in FEM modeling.

Let R_1 be the radius of the sapphire cylinders shown above and R_2 be the radius of the hole, which would fit between the cylinders and just touch each cylinder. The radius ratio of R_2 / R_1 is 0.155 by simple geometrical considerations[16]. The upper bound of the hole size in between the sapphire cylinders is approximately 5.4μm because sintering will tend to decrease the size of the hole. Using the radius ratio above, this gives a radius of 35μm for the sapphire cylinders.

A platinum wire was used to hold the fibers together in order to preserve the specific arrangement seen in Figure 1. The fibers were cleaned again and then placed on a porous high purity alumina plate and inserted into a 1600°C furnace for 12 hours. The fibers were then removed from the furnace and allowed to cool to 27°C. The fibers were then mounted in a polishing fixture and bonded using a thermoplastic adhesive. Figure 2 shows a micrograph taken in reflected light to highlight the structural aspects of the fiber[25]. This image is the basis of the 6-ring structure in the FEM modeling work.

Figure 2. Micrograph of 6-ring structure taken in reflected light.

The sapphire rods in this study were grown with the C-axis of the crystal extending along the length of the fiber. The hexagonal crystal structure of single crystal sapphire has two axes perpendicular to the c-axis, which are equivalent and referred to as the a-axis. The cross-section micrograph shown in Figure 2 is taken perpendicular to the c-axis of each of the fiber rods.

Figure 3 shows the transmitted light under white light illumination from the backside of the fiber[25].

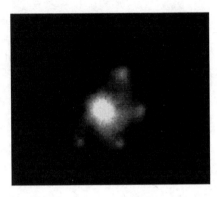

Figure 3. Micrograph of transmitted light in the 6-ring structure under white light illumination from the backside of the fiber.

Figure 3 shows a significant amount of the light propagating in the central rod of the fiber. This is consistent with confinement in the core due to the lowering of the average effective refractive index created by the holes in the PCF structure. Very faint light spots can be seen surrounding the central fiber, which correspond to the center positions of the sapphire cylinders surrounding the central rod in the 6-ring sapphire arrangement. The sapphire PCF shown was then heated up to 1600°C to demonstrate the high temperature capability of these single crystal sapphire fibers.

The experimental results have been published in a previous paper[25]. The objective of this paper was to model the propagation characteristics of a single rod of single crystal sapphire as well as the 6-ring around a central rod of single crystal sapphire. This would permit the resulting changes in optical characteristics of each fiber to be determined. This process would then allow for easy modification of the current size and positioning of the fiber rods in order to further analyze what effect this will have on the confinement loss.

FEM ANALYSIS

Numerical modeling is a key component while creating an optimal fiber design. Comsol Multiphysics uses high-order vectorial elements composed of both an automatic and iterative grid refinement calculator for optimum error estimation. This process is a necessity when accurate sapphire-air interfaces need to be sampled during the meshing stage. The high contrast between the 6-ring structure and the smaller air-hole sections puts a large burden on the computer's memory because of fast far-field variations. Add to this the fact that the electric field's normal component will become discontinuous and you get quite complex equations. A linear solver is typically used to compute the Maxwell's equations involved in the FEM discretization. To compensate for this, Comsol Multiphysics uses a direct linear system solver (UMFPACK) for its modal analysis.

Materials selection is the first main topic in sapphire photonic crystal modeling. We start with two separate fibers as described previously, which are scaled to the dimensions being demonstrated experimentally, as seen in Figure 3. The single crystal sapphire rods in these models have a diameter of 73μm. The refractive index of the air region is set to n = 1.0 and the sapphire region (α-Al$_2$O$_3$) is set to n = 1.74618. The air (blue) and sapphire (grey) regions can be seen in Figure 4. All models are solved for a free space wavelength of 1.55μm.

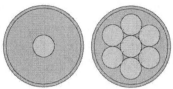

Figure 4. The air (blue) region is set to n = 1.0 and sapphire (grey) (α-Al$_2$O$_3$) is set to n = 1.74618.

The outer diameter of the fiber is set as a perfectly matched layer (PML). The outer diameter for the PML layer is set to 140μm with an inner PML diameter of 130μm. This utilizes a reflectionless outer layer used to absorb all outgoing waves. The PML is reflectionless for any frequency, thus absorbing all outgoing radiation[26]. No modes are expected in this outer air layer. Using a PML region also has the added benefit of saving valuable memory space.

Confinement loss, L_c, is a commonly used metric for measuring how an optically modeled fiber will perform under optimal operating conditions. It is directly related to the imaginary portion of the propagation constant[27], γ through the equation:

$$L_c = 8.686*\alpha \tag{1}$$

where α is the attenuation constant. This equation gives the loss[28] in units of dB/m. The propagation constant is defined as:

$$\gamma = \alpha + j\beta \tag{2}$$

where β is the phase constant. Maxwell's equations for a vector wave with an anisotropic PML is defined as:

$$\nabla \times ([s]^{-1}\nabla \times E) - k_0^2 n^2 [s]E = 0 \tag{3}$$

This equation is obtained from the uniformity of the electric field, which gives us:

$$E\ (x,\ y,\ z) = e(x,\ y)\exp(-\gamma z) \tag{4}$$

The next step in the modeling process is the refinement of a mesh. The accuracy of the output data is directly related to how precise the mesh is for each model. The mesh is limited by the memory of the computer that will be solving the boundary equations. Figure 5 below shows a mesh consisting of 24692 elements in the single rod case and 9508 elements for the 6-ring case. Both were computed on a 12Gb Macintosh I7 running Comsol 4.0a with a solution time of 875 seconds.

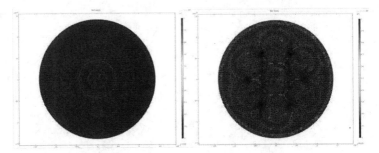

Figure 5. 2-D Comsol mesh in both the single rod (left) and the 6-ring case (right).

The boundary conditions for each fiber must now be setup. A perfect electrical conductor setting was used for the outer edge, while continuity was set inside the cylindrical PML. Comsol is then solved using a direct linear equation solver (UMFPACK) near the effective mode index of the single crystal sapphire (1.74618 at 1.55µm). Figure 6 below is the resultant fundamental hybrid mode for the single rod (left) and the 6-ring case (right). This is as expected for the optical power to be concentrated directly in the center of the core of both the single rod and the 6-ring central rod of both fibers. The single rod case has an effective mode index of 1.746109 with a corresponding confinement loss L_c= 2.0166e-8 dB/km. The 6-ring case also has an effective mode index of 1.746109 with a larger corresponding confinement loss L_c= 1.3933e-6 dB/km.

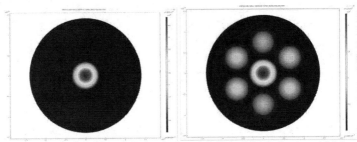

Figure 6. 2-D Comsol mesh in both the single rod (left) and the 6-ring case (right).

Figure 7 shows the higher order modes for the single rod case at a free space wavelength of 1.55µm. They consist (starting clockwise from top-left) of the LP_{11} mode with an effective mode index of 1.745999 and a corresponding confinement loss L_c= 3.9775e-8 dB/km, LP_{21} mode with an effective mode index of 1.745855 and a corresponding confinement loss L_c= 2.3718e-9 dB/km, LP_{02} mode with an effective mode index of 1.745804 and a corresponding confinement loss L_c= 2.7683e-9 dB/km, and the LP_{31} mode with an effective mode index of 1.745678 and a corresponding confinement loss L_c= 1.1347e-8 dB/km.

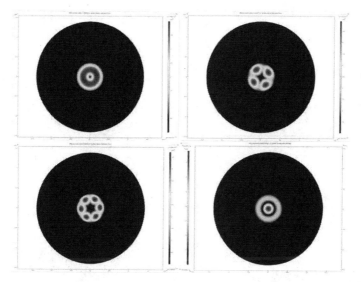

Figure 7. Higher order modes for the single rod case: (starting clockwise from top-left) LP_{11}, LP_{21}, LP_{02}, and LP_{31}.

Figure 8 shows the higher order modes for the 6-ring case at a free space wavelength of 1.55μm. They consist (starting clockwise from top-left) of the LP_{11} mode with an effective mode index of 1.746001 and a corresponding confinement loss L_c= 2.3942e-6 dB/km, LP_{21} mode with an effective mode index of 1.745858 and a corresponding confinement loss L_c= 8.2314e-6 dB/km, LP_{02} mode with an effective mode index of 1.745807 and a corresponding confinement loss L_c= 9.1295e-6 dB/km, and the LP_{31} mode with an effective mode index of 1.745686 and a corresponding confinement loss L_c= 6.7056e-6 dB/km.

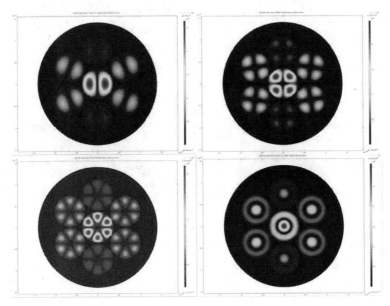

Figure 8. Higher order modes for the 6-ring case: (starting clockwise from top-left) LP_{11}, LP_{21}, LP_{02}, and LP_{31}.

CONCLUSION

A sapphire photonic crystal fiber has been presented and modeled using a multi-physics modeling program. Although this type of structure has been reported in silica-based fibers, this represents the first time that a photonic crystal fiber has been fabricated and modeled in single crystal sapphire. The single rod fiber as well as the 6-ring around a central rod fiber were both modeled to see if resulting changes in optical characteristics of each fiber could be determined. This process allows for easy modification of the current size and positioning of the fiber rods in order to further analyze what effect this will have on the confinement loss. The results also show that the confinement loss is directly related to which mode is propagating through the fiber.

ACKNOWLEDGEMENTS

The authors would like to gratefully acknowledge funding for this work from DOE NETL under contract number FC26-05NT42441.

REFERENCES

[1] T.A. Birks, J.C. Knight, and P.St.J. Russell, "Endlessly single-mode photonic crystal fiber," *Journal of Optics Letters* **22**, 961-963 (1997).

[2] J.C. Knight, T.A. Birks, P.J. Russell, and J.P. de Sandro, "Properties of photonic crystal fiber and the effective index model," *Journal of the Optical Society of America* **15**, 748-752 (1998).

[3] J.C. Knight, T.A. Birks, P.St.J. Russell, and D.M. Atkin, "All-silica single-mode optical fiber with photonic crystal cladding," *Journal of Optics Letters* **21**, 1547-1549 (1996).

[4] T. M. Monro, P. J. Bennett, N. G. R. Broderick, and D. J. Richardson, Opt. Lett. 25, 206 (2000).

[5] B. J. Mangan, J. C. Knight, T. A. Birks, and P. St. J. Russell, in Digest of Conference on Lasers and Electro-Optics (CLEO), Optical Society of America, Washington, D.C., 559 – 560 (1999).

[6] D. Kominsky, G. Pickrell and R. Stolen, "Generation of random-hole optical fiber," *Journal of Optics Letters* **28**, 1409-1411 (2003).

[7] G. Pickrell, D. Kominsky, R. Stolen, F. Ellis, J. Kim, A. Saffaai-Jazi, and A. Wang, "Microstructural analysis of random hole optical fibers," *Photonics Technology Letters* **16**(2), 491-493 (2004).

[8] T. Nasilowski, R.Kotynski, F. Berghman, H. Thienpont. "Photonic Crystal Fibers – state of the art and future perspectives," Lightguides and their applications II, Proc. SPIE **5576**, 1-12 (2004).

[9] B. Temelkuran, S.D. Hart, G. Benoit, J.D. Joannopoulos and Y. Fink. " Wavelength-scalable hollow optical fibers with large photonics bandgaps for CO_2 laser transmission," *Nature* **420**, 650-653 (2002).

[10] M. Korwin-Pawlowski, G. Pickrell, P. Mikulic, "Long Period Gratings on Random Hole Optical Fibers and Microstructured Disordered Fibers," Proc. SPIE **6767**, art. no. 12.735559 (2007).

[11] G. Pickrell, W. Peng and A. Wang, "Random-hole optical fiber evanescent-wave gas sensing," *Journal of Optics Letters* **29**, 1476-1478 (2004).

[12]B. Alfeeli, G. Pickrell, M. Garland and A. Wang, "Behavior of Random Hole Optical Fibers Under Gamma Ray Irradiation and its Potential Use in Radiation Sensing Applications," *Sensors* 7(5), 676-688 (2007).

[13] G. Pickrell, C. Ma, and A. Wang, "Bending-induced optical loss in random-hole optical fibers," *Journal of Optics Letters* **33**, 1443-1445 (2008).

[14] P.St.J. Russell, " Photonic crystal fibers," Science **299**, 358-362 (2003)

[15] Vincent Pureur, Géraud Bouwmans, Mathias Perrin, Yves Quiquempois, and Marc Douay, "Impact of Transversal Defects on Confinement Loss of an All-Solid 2-D Photonic-Bandgap Fiber," J. Lightwave Technol. **25**, 3589-3596 (2007)

[16] W.D. Kingery, H.K. Bowen and D.R. Uhlmann, *Introduction to Ceramics*, John Wiley and Sons, New York, NY (1976).

[17] G. R. Pickrell "High-temperature alkali corrosion kinetics of low-expansion ceramics (alkali corrosion)," Virginia Polytechnic Institute & State University, Blacksburg, VA, Ph.D. dissertation (1994).

[18] R. S. Feigelson, Pulling optical fibers, Journal of Crystal Growth, Volume 79, Issues 1-3, Part 2, Proceedings of the Eighth International Conference on Crystal Growth, 2 December 1986, Pages 669-680, ISSN 0022-0248, DOI: 10.1016/0022-0248(86)90535-X.

[19] Ke Wang, Liejia Qian, Hang Luo, Peng Yuan, and Heyuan Zhu, "Ultrabroad supercontinuum generation by femtosecond dual-wavelength pumping in sapphire," Opt. Express 14, 6366-6371 (2006).

[20] H. Xiao, J. Deng, G. Pickrell, R. G. May, and A. Wang, "Single-Crystal Sapphire Fiber-Based Strain Sensor for High-Temperature Applications," *Journal of Lightwave Technology* **21**, 2276 (2003).

[21] A. Wang, S. Gollapudi, R.G. May, K.A. Murphy and R.O. Claus, "Advances in sapphire-fiber-based interferometric sensors," *Journal of Optics Letters* **17**, 21 (1992).

[22] A. Wang, S. Gollapudi, K. A. Murphy, R. G. May, and R. O. Claus, "Sapphire-fiber-based intrinsic Fabry-Perot interferometer," *Journal of Optics Letters* **17**, 14 (1992).

[23] A. Wang, P. Zhang, K. A. Murphy, and R. O. Claus, "Sapphire optical fiber-based polarimetric sensors for high temperature applications," *SPIE, Smart Materials and Structures 1994: Smart Sensing, Processing, and Instrumentation,* Feb., (1994).

[24] A. Wang, G. Z. Wang, K. A. Murphy, and R. O. Claus, "Birefringence-balanced polarization-modulation sapphire optical fiber sensor," *Journal of Optics Letters* **17**, 1391 (1992).

[25] Neal Pfeiffenberger, Gary Pickrell, Karen Kokal, and Anbo Wang, "Sapphire photonic crystal fibers," Opt. Eng. 49, 090501 (2010), DOI:10.1117/1.3483908

[26] Jianming Jin, "The Finite Element Method in Electromagnetics," Second Edition, John Wiley & Sons, Inc, New York 2002.

[27] K. Saitoh and M. Koshiba, "Full-vectorial imaginary-distance beam propagation method based on finite element scheme: Application to photonic crystal fibers," IEEE J. Quantum Electron. **38**, 927-933 (2002).

[28] Kunimasa Saitoh and Masanori Koshiba, "Leakage loss and group velocity dispersion in air-core photonic bandgap fibers," Opt. Express **11**, 3100-3109 (2003)

MAGNETICALLY-DRIVEN RELEASE MEDIA COMPRISING OF CARBON NANOTUBE-NICKEL/NICKEL OXIDE CORE/SHELL NANOPARTICLE HETEROSTRUCTURES INCORPORATED IN POLYVINYL ALCOHOL

Wenwu Shi, Nitin Chopra*

Metallurgical and Materials Engineering, Center for Materials for Information Technology (MINT), The University of Alabama, Tuscaloosa, AL 35401 USA
*Corresponding author: Tel: 205-348-4153; Fax: 205-348-2164; E-mail: nchopra@eng.ua.edu

ABSTRACT

We report a simple synthetic approach for carbon nanotube (CNT)-Ni/NiO core/shell nanoparticles heterostructures (CNC heterostructures) in a single-step direct nucleation and air oxidation of Ni nanoparticles on multi-walled CNTs. Detailed characterization (SEM and TEM) of CNC heterostructure was carried out and as a next step these heterostructures were successfully incorporated into a polyvinyl alcohol (PVA) hydrogel to result in a magnetic hydrogel (CNC heterostructure-PVA hydrogel), which could be deflected by 60° under a 0.2 T permanent magnet. In addition, CNC heterostructures were loaded (physically adsorbed) with L-histidine and then incorporated within the PVA hydrogel. This unique nanosystem was studied for the release of loaded L-histidine, with and without the external magnetic field. It was observed that the amount of histidine released from the hydrogel was increased by 12.9% under 0.2 T magnetic field as compared to the release without any magnetic field. Such a unique hybrid hydrogel incorporating magnetic CNC heterostructures hold great promise for smart drug delivery nanosystems.

INTRODUCTION

The main focus in the area of thin film heterostructures has been modulating and improving device characteristics (e.g., band gap energies and charge carrier mobilities) through composition variations in epitaxially grown thin films with close interfacial lattice match.[1-4] This has led to numerous advances in light emitting diodes, solar cells, and transistors.[5-7] Heterostructures are not only critical for electronics and optical devices but also for novel chemical and biological sensors and drug delivery. Furthermore, with the emergence of nanomaterials[8-11] and advances in growth and characterization tools, nanotechnology research is gradually shifting from simple single component nanostructures to complex multi-component nanoscale heterostructures.[12-18] The latter can be comprised of a combination of nanostructures of different or same material(s). Among these, carbon nanotubes (CNT)-based heterostructures, especially CNTs coated with nanoparticles are of great importance because of their unique properties and multi-functionality.[19,20] However, controlled synthesis of these heterostructures requires the development of simple and facile synthetic routes with no covalent linkages, yet resulting in robust CNT-nanoparticle heterostructures. CNT-nanoparticle heterostructures have been commonly fabricated using covalent chemistry-based linking of nanoparticles with CNTs,[20] which severely limits the choice of nanoparticle materials. The challenge is to attain a simple fabrication route that results in heterostructured 1-D nanostructures with minimal contamination and uniform loading of one nanostructure (e.g., nanoparticles) on the other (e.g., CNTs). In this regard, a recent report[19] by the authors shows the formation of CNT-Ni/NiO core/shell nanoparticles heterostructures and their incorporation into PVA hydrogel (CNC-PVA hydrogel). This was a direct nucleation approach, where nanoparticles were nucleated on high curvature CNTs resulting in a very uniform and tightly packed coating of core/shell nanoparticles. Later, these heterostructures were incorporated in a PVA hydrogel and studied for selectively concentrating proteins in a solution. Motivated by the results of this work, here we report our initial study on developing magnetically actuating CNC-PVA hydrogel as well as their enhanced delivery response under an applied magnetic field. These preliminary results are very promising for developing smart chemical and biological separation or delivery nanosystems.

EXPERIMENTAL

Materials and Methods: Nickel acetate tetrahydrate, oleylamine, trioctylphosphine (TOP) and ferrocene were purchased from Sigma-Aldrich (St. Louis, MO). Tri-n-octylphosphine oxide (TOPO) was purchased from Alfa aesar (Ward Hill, MA). Ethanol, acetone, xylenes, hexane, and silicone oil were bought from Fisher Scientific (Pittsburgh, PA). All chemicals were directly used without purification. Labnet centrifuge (Edison, NJ) was used to clean and separate heterostructures. Wet samples were dried inside VWR vacuum oven (West Chester, PA).

Synthesis of CNTs: Multi-walled CNTs were synthesized by chemical vapor deposition (CVD) using ferrocene and xylene as catalyst and carbon source, respectively. Ferrocene/xylene liquid mixtures were injected through syringe injector into a pre-heated zone and subsequently transported into reaction zone (~675 °C) inside the quartz tube furnace with H_2/Ar as oxygen scavenger as well as carrier gas. The reaction continued for 2 hours and then the furnace was cooled down slowly. CNTs (black powder) were collected from the inner walls of the quartz and characterized later by electron microscopy.

Synthesis of CNT-Nickel/Nickel oxide core/shell nanoparticle (CNC) heterostructures: CNC heterostructures were fabricated by a simple technique using direct nucleation of Ni nanoparticles on CNTs. Exposing heterostructures to the air resulted in the formation of NiO shell around the Ni core. Typically, 1.00 g nickel acetate tetrahydrate, 7 mL oleylamine and 0.01 g CNTs were mixed in a beaker and magnetically stirred for 10 min followed by 30 min ultrasonication. The black mixture was transferred into a three-neck round-bottom flask kept inside silicone oil bath. The mixture was held at 90 - 95°C under N_2 atmosphere for 40 min. Subsequently, 1.50 g TOPO and 1 mL TOP were added to the hot solution. The temperature was gradually increased from 90 to 250 °C at ~10 °C/min and held at 250 °C for 30 min. After the reaction, the flask was allowed to cool in the air and washed several times in ethanol, hexane, and acetone. The obtained powder was dried overnight in a vacuum oven at 80 °C and stored in a glass vial in air. In order to eliminate remaining stabilizers such as oleylamine, TOPO, and TOP, ligand exchange reaction with CNC heterostructures with imidazole/chloroform (0.5 g/mL) inside capped glass vial was performed.

Preparation of L-histidine loaded CNC heterostructure-PVA hydrogel: 0.08 g of cleaned and dried CNC heterostructures were physically adsorbed with L-histindine in DI water and incubated for 1 h. On the other hand, a PVA solution (in DI water) was prepared by mixing PVA and Poly-ethyl glycol (PEG-600) and refluxed at 95 °C with stirring for 1 hour. In the next step, the prepared CNC heterostructures loaded with L-histidine were mixed in the polymer solution. The resulting sticky solution was then poured into a 2" petri dish and frozen at -20 °C for 24 hours. The hydrogel was thawed and dried inside the hood at ambient temperature. This process lead to the formation of CNC heterostructure-PVA hydrogel loaded with L-histidine. The magnetic character of these hydrogels was also tested by actuating them under a 0.2 T permanent magnet.

Release of L-histidine: Multifunctional CNC heterostructure-PVA hydrogel loaded with L-histidine was observed for its release behavior with and without external magnetic field. For all the experiments, 50 mg dried hydrogel was immersed in 50 mL DI water at room temperature and gently shook. For every 5 minutes, 3 mL solution was sampled and tested for UV absorbance peak of L-histidine (213 nm). This release was also studied by keeping a 0.2 T permanent magnet outside the beaker containing the CNC heterostructure-PVA hydrogel and measuring the UV absorbance of the released L-histidine.

Characterization: Microscopic characterizations were performed using Field Emission Scanning Electron Microscope (FE-SEM, JEOL-7000) and transmission electron microscopy (HR-TEM, Tecnai FEI-20). UV-vis spectra were collected using UV-Vis spectrometer (USB-4000, Ocean Optics).

RESULTS AND DISCUSSION

Heterostructures based on CNTs and nanoparticles are of great importance for their unique chemical properties and surface functionality. Simple and effective techniques that produce large amounts of such heterostructure are highly desired. Figure 1A shows the CNTs formed in our CVD growth process. These CNTs have an average diameter of ~ 42.7±12.3 nm and length of ~ 26 μm. As shown by TEM image (Figure 1B), these CNTs have an inner core of ~ 5 nm with iron catalyst nanoparticle stuck within the core and there is also some extent of stress marks (striations, arrows showing in Figure 1B) seen on the walls of the CNTs. These stress marks can be due to the growth process as well as natural cooling (thermal gradients) of the CNTs after the CVD furnace is switched off.[21,22] In addition, it can be seen from the TEM image (Figure 1B) that these CNTs have minimal amount of amorphous carbon coating on their surface. This makes them more suitable for direct nucleation of nanoparticles in a chemical synthetic route. As described, precursor nickel salt mixed with a solution containing as-produced CNTs and stabilizers were reacted together. This resulted in direct nucleation of Ni nanoparticles. Further exposure to air led to the formation of Ni/NiO core/shell nanoparticles on the surface of CNTs as shown by the SEM and TEM images (Figure 1C and D). The average diameter of nanoparticles was observed to be ~ 11.8±1.7 nm. Once these heterostructures were prepared, they were incorporated into a PVA hydrogel resulting in magnetic CNC heterostructure-PVA hydrogel as shown in Figure 2A. When a 0.2 T permanent magnet is brought near the hybrid hydrogel saturated with water, a deflection or actuation angle of 60° is observed for the latter. This is a large deflection as compared to previously reported CNT-based hydrogel that required incorporation of sodium ions to actuate them in an external electric field.[23] Thus, the novel CNC heterostructure-PVA hydrogel reported here eliminates complex actuating mechanism by simply introducing magnetic functionality into the multi-component nanosystem through Ni/NiO core/shell nanoparticles.

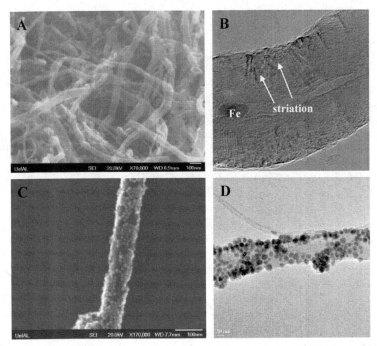

Figure 1. SEM image of A) as produced CNTs, high resolution TEM image of B) as produced CNT, SEM image of C) CNC heterostructure, and TEM image of D) CNC

Finally the magnetic CNC heterostructures were loaded with L-histidine and incorporated into the PVA hydrogel. L-histidine was considered as a model molecule to indicate the potential of these hybrid hydrogels for drug delivery nanosystems or magnetically-driven release media. It was observed that L-histidine release from the hybrid hydrogel was increased by 12.9% when a 0.2 T permanent magnet was kept near the hydrogel as compared to the release without magnet (Figure 2B). Authors propose that the reason for the increased release under magnetic field was due to higher degree of stretching of hydrogel that further expanded the pore size of the hybrid hydrogel away from the magnet (as the magnet is placed on one side of the hydrogel). This PVA pore stretching is directly related to the magnetic force exerted on the CNC heterostructures incorporated inside PVA hydrogel. This is a unique observation and further studies are underway in the authors' laboratory to evaluate the release behavior and effects of the magnetic field.

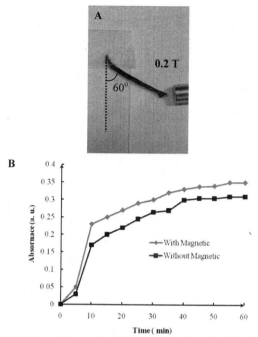

Figure 2. A) Deflection of magnetic CNC-PVA hydrogel under the influence of 0.2 T permanent magnet, B) release of L-histidine (physically adsorbed onto CNC heterostructures incorporated into PVA hydrogel) with and without magnetic field.

CONCLUSIONS

A simple and single-step chemical synthetic approach is reported here for the direct nucleation of Ni/NiO core/shell nanoparticles on CNTs resulting in novel and multi-functional CNC heterostructures. This process resulted in a uniform coating of the nanoparticles and also rendered magnetic character to the CNC heterostructures. Finally, CNC heterostructures were incorporated into PVA hydrogel (CNC heterostructure-PVA hydrogel) that could be easily deflected by large angles (60°) under a 0.2 T magnetic field. An interesting application of CNC-PVA hydrogel as a magnetically-driven release media was demonstrated by loading CNC heterostructures with L-histidine and then incorporating them into PVA hydrogel. The release of physically adsorbed histidine was studied with and without magnetic field. It was observed that when the loaded hydrogel was immersed in the DI water and a magnet is brought near it, the release of histidine increased by 12.9% as compared to the release without a magnet. This proof-of-concept release experiments indicate the strong potential of such hybrid hydrogels for smart drug delivery devices.

ACKNOWLEDGEMENTS

The authors thank the University of Alabama (Department of Metallurgical and Materials Engineering and the Office of sponsored programs), The University of Alabama's Research Grant Committee Award, start-up funds from the University of Alabama, and National Science Foundation (NSF award #: 0925445) for supporting this work. The authors also thank the Central Analytical Facility (CAF) for electron microscopy equipment (NSF-MRI funded) and the financial support covering the instrument time, the MINT Center for providing infrastructure support such as clean room facility and various equipments, and Mr. Rich Martens, Mr. Rob Holler, and Mr. Johnny Goodwin for providing training on SEM and TEM. The authors thank Dr. Shweta Kapoor for proof reading the manuscript.

REFERENCES

[1] Z.I. Alferov, Classical heterostructures paved the way, *III-Vs Review*, **11**, 26-31 (1998).

[2] S. M. Sze, Physics of semiconductor devices, John Wiley and Sons (Canada), 1981.

[3] C. H. Ahn, K. M. Rabe, J. M. Triscone, Ferroelectricity at the Nanoscale: Local Polarization in Oxide Thin Films and Heterostructures, *Science*, **303**, 488-491 (2004).

[4] Y. I. Alivov, E. V. Kalinina, A. E. Cheenkov, D. C. Look, B. M. Ataev, A. K. Omaev, M. V. Chukichev, D. M. Bagnall, Fabrication and characterization of n-ZnO/p-AlGaN heterojunction light-emitting diodes on 6H-SiC substrates, *Appl. Phys. Lett.*, **83**, 4719-4721 (2003).

[5] D. K. Hwang, S. H. Kang, J. H. Lim, E. J. Yang, J. Y. Oh, J. H. Yang, S. J. Park, p-ZnO/n-GaN heterostructure ZnO light-emitting diodes, *Appl. Phys. Lett.*, **86**, 222101/1-3(2005).

[6] J. Heber, Enter the oxides, *Nature*, **459**, 28-30 (2009).

[7] C. K. Wang, Y. Z. Chiou, S. J. Chang, Y. K. Su, B. R. Huang, T. K. Lin, S. C. Chen, AlGaN/GaN metal-oxide semiconductor heterostructure field-effect transistor with photo-chemical-vapor deposition SiO_2 gate oxide, *J. Electron. Mater.*, **32**, 407-410 (2007).

[8] Y. Wu, H. Yan, M. Huang, B. Messer, J. H. Song, P. Yang, Inorganic Semiconductor Nanowires: Rational Growth, Assembly, and Novel Properties, *Chem. Eur. J.*, **8**, 1260-1268 (2002).

[9] Y. Wu, H. Yan, P. Yang, Semiconductor Nanowire Array: Potential Substrates for Photocatalysis and Photovoltaics, *Top. Catat.*, **19**, 197-202 (2002).

[10] F. Patolsky, G. Zheng, C. M. Lieber, Nanowire-based biosensors, *Anal. Chem.*, **78**(13), 4260-4269 (2006).

[11] N. Chopra, V. Gavalas, B.J. Hinds, L.G. Bachas, Functional one-dimensional nanomaterials: Applications in nanoscale biosensors, *Anal. Lett.*, **40**, 2067-2096 (2007).

[12] Y. Li, F. Qian, J. Xiang, C. M. Lieber, Nanowire electronics and optoelectronics, *Mater. Today*, **9**, 18-27 (2006).

[13] A. K.Salem, J. Chao, K. W. Leong, P. C. Searson, Receptor-Mediated Self-Assembly of Multi-Component Magnetic Nanowires, *Adv. Mater.*, **16**, 268-271 (2004).

[14] K. A. Dick, S. Kodambaka, M. C. Reuter, K. Deppert, L. Samuelson, W. Seifert, L. R. Wallenberg, F. M. Ross, The Morphology of Axial and Branched Nanowire Heterostructures, *Nano Lett.*, **7**, 1817-1822 (2007).

[15] N. Chopra, L. G. Bachas, M. Knecht, Fabrication and Biofunctionalization of Carbon-Encapsulated Au Nanoparticles, *Chem. Mater.*, **21**, 1176-1178 (2009).

[16] N. Chopra, M. Majumder, B. J. Hinds, Bi-functional carbon nanotubes by sidewall protection, *Adv. Funct. Mater.*, **15**(5), 858-864 (2005).

[17] W. Shi, N. Chopra, Surfactant-free synthesis of novel copper oxide (CuO) nanowire–cobalt oxide (Co_3O_4) nanoparticle heterostructures and their morphological control, *J. Nanopart. Res.*, 2010, In Press (DOI: 10.1007/s11051-010-0086-0).

[18] N. Chopra, Multi-functional and Multi-component Heterostructured One-Dimensional Nanostructures: Advances in Growth, Characterization, and Applications, Mater. Technol., **25**, 212-230 (2010).

[19] W. Shi, K. Crews, N. Chopra, Multi-Component and Hybrid Hydrogels Comprised of Carbon Nanotube-Nickel/Nickel Oxide Core/Shell Nanoparticle Heterostructures Incorporated in Polyvinyl Alcohol, Mater. Technol., **25**, 149-157 (2010).

[20] X. H. Peng, J. Y. Chen, J. A. Misewich, S. S. Wong, Carbon nanotube–nanocrystal heterostructures, *Chem. Soc. Rev.*, **38**, 1076-1098 (2009).

[21] N. Chopra, P. D. Kichambare, R. Andrews, B. J. Hinds, Control of multiwalled carbon nanotubes diameter by selective growth on the exposed edge of a thin film multilayer structure, *Nano Lett.*, **2**(10), pp 1177–1181 (2002).

[22] Meyyappan, M. Carbon nanotubes: Science and applications. CRC Press LLC, Boca Raton, FL, 2005.

[23] J. Shi, Z. X. Guo, B. Zhan, H. Luo, Y. Li, D. Zhu, Actuator based on MWNT/PVA hydrogels, *J. Phys. Chem. B*, **109** (31), 14789–14791 (2009).

SINGLE-WALLED CARBON NANOTUBE DISPERSION STRUCTURES FOR IMPROVED ENERGY DENSITY IN SUPERCAPACITORS

Joshua J. Moore, Dr. John Z. Wen
University of Waterloo, 200 University Avenue West
Waterloo, Ontario, Canada N2L 3G1
jje2moor@uwaterloo.ca

ABSTRACT

Single walled carbon nanotubes (SWCNT) were dispersed into aqueous solutions without surfactant by functionalizing the SWCNT with carboxyl groups. SWCNT coated electrodes were then prepared from the SWCNT aqueous solutions using drop coating, high voltage electro-spinning (HVES), and electrophoretic deposition (EPD) methods using Ni and SST foils as the current collectors. Optical and scanning electron microscope images were used to evaluate the SWCNT dispersion quality of the various methods. An EPD process was established which reliably produced uniform SWCNT nanoporous networks on Ni and SST foils. The resultant SWCNT coated electrodes were characterized using cyclic voltammetry to evaluate their capacitance. A direct correlation between EPD processing time and capacitance was discovered. The addition of the SWCNT nanoporous network to the SST electrode resulted in an increase in capacitance of 45 times the capacitance of the uncoated SST.

INTRODUCTION

As energy dependence moves away from fossil fuels and consumable energy sources the need for energy storage devices with larger storage capacity and higher charge rates becomes more and more critical to the portable energy market[1]. There is a technology gap between the large storage capacity of batteries and the fast charge and discharge of capacitors. New and innovative materials and processes are being developed in an effort in a fill this gap[2]. Supercapacitors do not rely on chemical reactions for charge transfer and thus are not limited in speed or useful life by the chemical changes[3]. The charge from a supercapacitor is created by the movement of ions in an electrolyte to opposite electrodes of the supercapacitor cell. From an electrode standpoint the amount of charge is primarily dependent on the effective surface area sites available for ion attraction and the separation distance between the electrodes as shown in the following equation[3]:

$$C = \epsilon A/d \tag{1}$$

Where A is the surface area of an electrode pair, d is the distance between them and ϵ is the permittivity relating to the electrolyte.

However, new research has shown that the pore size in porous materials can play an important role in the amount of charge an electrode can store[4-5]. The research shows that there is a critical pore size less than 1nm in diameter which can provide the largest charge storage capability. Typical porous materials for use in supercapacitors such as Nickel or Carbon foams and fabrics have pore sizes much larger than this critical nanopore size which can be in the order of micrometers[6]. Control of pore sizes is critical to optimize the ion trapping sites and increase the stored charge on supercapacitor electrodes. A process is needed which is repeatable and can uniformly deposit as produced SWCNTs over large areas for high volume production of SWCNT electrodes while producing uniform nano pores to create maximum charge capacity. Investigation of dispersed single wall carbon nanotubes (SWCNT) to form nanopores for electrolyte ion trapping in supercapacitors enables tailoring of structures for optimum storage capacity. Various SWCNT structures, as formed through dispersion of SWCNT solutions, and their effect on the energy density of supercapacitors are investigated here. Different SWCNT dispersion techniques such as drop coating, high voltage electro-spinning and electrophoretic deposition[7-8,13] were investigated to form various nano-structures, however, the primary focus of this paper is the use of electrophoretic deposition (EPD) to create the nanoporous SWCNT networks. Understanding and optimizing the SWCNT structures to produce large areas of uniform nanopores will increase the energy density of supercapacitors and help to bridge the gap between high energy density and high power density storage devices.

EXPERIMENTAL PROCEDURES

Purified SWCNTs obtained from Nano-C were acid functionalized using a mixture of concentrated nitric (HN0$_3$) and sulfuric (H$_2$SO$_4$) acid in a ratio of 1:3 respectfully (Sigma-Aldrich). 10mL of the acid mixture was added to 250mg of SWCNT powder for 30min to complete the carboxyl functionalization[9-10]. The acid mixture adds a –COOH functional group to defect sites on the SWCNTs[11-12]. This negatively polarizes the SWCNTs and allows them to disperse well in pure H$_2$O without the use of additional surfactants[13]. The addition of surfactants has been shown to negatively affect the performance of CNTs and it is difficult to remove the surfactant from the CNT network after processing[14]. After acid functionalizing, the SWCNT acid mixture was diluted with de-ionized water washed through a filter to remove the acidity from the CNTs. Once the CNTs had been neutralized the functionalized SWCNTs were washed from the filter into 500ml of di-water to produce a SWCNT aqueous solution with a SWCNT concentration of approximately 0.5mg/mL H$_2$O. The solution was then placed in a Branson 5210 Ultrasonic cleaner for 30min at 22°C. Prior to EPD the solution is placed in ultrasonic bath for an additional 5min and 10mL of prepared solution is removed and centrifuged for 30min at 4000RPM to remove agglomerated SWCNT bundles from the solution (Figure 1).

Figure 1: Functionalized SWCNT aqueous solutions

An EPD cell was fabricated with a working distance of 1cm between electrodes (Figure 2). Electrodes for the EPD cell also acted as the current collectors for the SWCNT supercapacitors. Two types of electrode materials were used, 0.127mm thick 99% pure Ni foil and 0.100mm thick stainless steel (SST) (70%Fe, 19%Cr, 11%Ni: wt%) from Alfa Aesar. The Ni and SST electrodes were cut, numbered, and weighed prior to deposition then installed into the EPD cell and lowered into the functionalized SWCNT aqueous solution to a depth of approximately 1cm. A DC power supply was used to apply 10-50V to the EPD cell for a time duration of 1-8 minutes. Over time the negatively charged functionalized SWCNTs moved towards the positive electrode and uniformly coated the surface. The SWCNT coated positive electrode was removed from the solution under applied voltage then disconnected from the voltage supply in air. The SWCNT coated electrodes were then dried in an oven at 80°C for 30min.

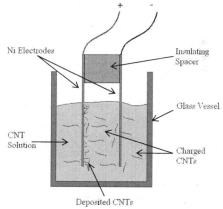

Figure 2: Schematic diagram of Carbon Nanotube (CNT) EPD cell

Drop coated and high voltage electro-spinning (HVES) samples were also prepared using the same centrifuged SWCNT aqueous solutions. For drop coating approximately 3-4mL of solution was applied to horizontal Ni and SST plates using a micro-pipette and allowed to dry in air for 24hrs before oven drying for 30min at 80°C. In high voltage electro-spinning Ni and SST electrodes were grounded and placed approximately 25cm below a dispensing needle containing 30mL of SWCNT solution. An 8kV potential is introduced between the dispensing needle and the electrodes and the solution is slowly ejected from the needle. Figures 3 and 4 below show schematic diagrams of the drop and HVES electrode preparation.

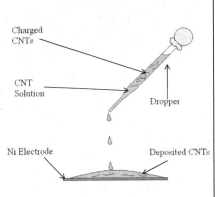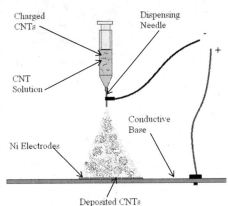

Figure 3: Schematic diagram of SWCNT drop coating process

Figure 4: Schematic diagram of HVES SWCNT coating process

Scanning electron microscope (SEM) images were gathered on a LEO 1530 FE-SEM, and were used to evaluate the uniformity and visualize the porosity of the coated SWCNT networks on the Ni and SST current collectors. Surface area and average pore size was also measured on the SWCNT coated electrodes using a NOVA 2200e BET tester. Cyclic voltammetry (CV) measurements were performed on a Princeton Applied Research Potentiostat / Galvanostat Model 273A, within a potential range of 0.0-1.0V at scan rates of 20, 50, and 100mV/s.

RESULTS AND DISCUSSION

Figure 5 below shows examples of the SWCNT coated Ni and SST electrodes after EDP, drop, and HVES coating methods. For both the drop and HVES coating methods it was found from optical, SEM images, and tape adhesion tests that the dispersion qualities were non-uniform and/or had poor adhesion to the Ni and SST electrode materials. As a result, the drop and HVES processing methods were determined not suitable for the dispersion of SWCNT on Ni or SST to create supercapacitor electrodes.

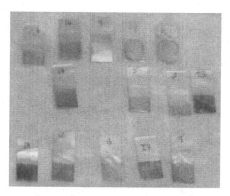

Figure 5: Ni and SST electrodes coated with SWCNTs using
EPD, drop, and HVES methods

The electrophoretic deposition method (EPD), however, showed promising results and thus was used as the primary process for electrode fabrication and characterization. SEM images of Ni electrodes coated with SWCNTs at various applied voltages between 10V and 50V and for various processing times between 1min and 8min confirmed the presence of a SWCNT nanoporous network. Figure 6 below shows the nanoporous network of an EPD SWCNT coated Ni electrode processed at 30V for 3min. The images show a uniform coating and an average pore size of approximately 20nm. BET surface area analysis was also used to confirm the average pore size for the 30V 3min EPD sample at 22.3nm for all pores less than 124.2nm.

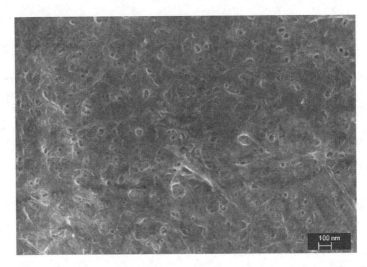

Figure 6: SEM image of well formed nanoporous SWCNT network on Ni electrode using EPD

SEM images of electrodes processed using the drop and HVES coating methods are also shown below (Figure 7-8) and confirmed what was observed optically in that the networks are poorly formed and not uniformly coated.

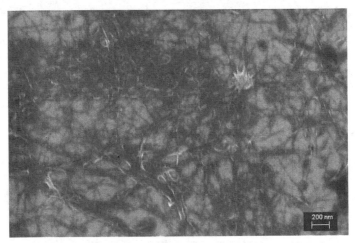

Figure 7: SEM image of poorly formed SWCNT network on Ni using HVES method

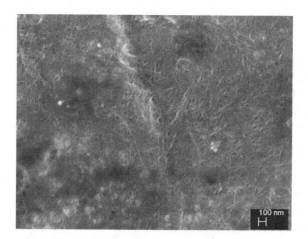

Figure 8: SEM image of poorly formed SWCNT network on Ni using drop coating method

SEM images of poor networks were also observed for EPD samples processed at low voltage 10 to 20V or had short processing times 1 to 2min. For these low voltages and short processing times the formation of the SWCNT networks was sparse and the overall coverage of the electrode was not uniform. Areas of bare Ni or bare SST could be readily observed. As expected, there appears to be a direct correlation between the EPD voltage and processing time and the SWCNT network coverage. This formation of the network can be observed visually from Figure 9 and Figure 10 below. SEM images of the 30V 3min processed electrode show the SWCNT network formation over a hole in the SWCNT coating. SWCNT bundles can be observed spanning the hole and forming the SWCNT network. It is hypothesized that given a longer processing time the hole on this sample would have completely filed in with SWCNTs.

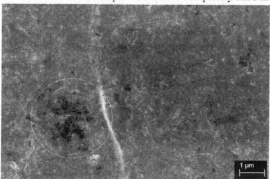

Figure 9: SEM image of SWCNT network forming across hole

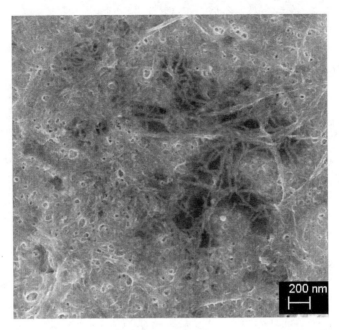

Figure 10: High magnification SEM image of SWCNT bundles spanning hole to form SWCNT network.

Through SEM analysis of the SWCNT Ni and SST electrodes processed at various EPD voltages 10-50V it was determined that an EPD voltage of 40V or higher consistently produced the most uniform SWCNT networks. At a constant EPD voltage of 40V, processing times of 0min, 3min, 5min, and 8min where used on SST electrodes to investigate the SWCNT network formation over time. Cyclic voltammetry (CV) curves were gathered to determine the effect of the EPD processing time on the capacitance of SWCNT coated SST samples. The optical images below show some of the uncoated and coated SST electrodes used to collect the CV curves (Figure 11).

 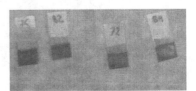

Figure 11: Optical images of uncoated SST electrodes (left) and SWCNT coated SST electrodes (right) using 40V EPD process

CV curves were performed on single processed electrodes using an Ag/AgCl reference electrode and a Pt counter electrode in a three electrode cell. The working electrolyte for the CV curves was 1M HSO_4. During the CV tests the bulk of the SST electrodes were masked with Teflon tape. Only the SWCNT area on one side, approximately $1cm^2$, was left exposed to the electrolyte. Figure 11-13 below show that there is a direct correlation between EPD processing time and the capacitance of the SWCNT coated SST electrodes.

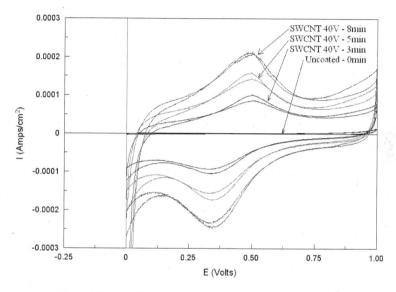

Figure 12: Cyclic Voltammetry curves of SWCNT coated SST
electrodes at 20mV/s showing effect of EPD processing time on capacitance

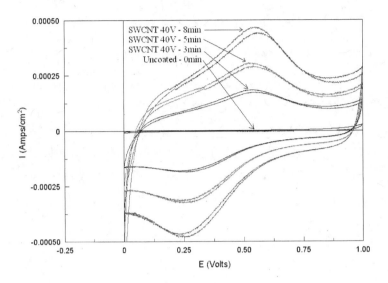

Figure 13: Cyclic Voltammetry curves of SWCNT coated SST electrodes at 50mV/s showing effect of EPD processing time on capacitance

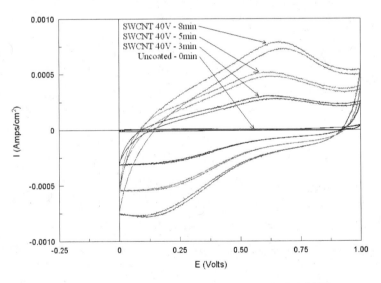

Figure 14: Cyclic Voltammetry curves of SWCNT coated SST electrodes at 100mV/s showing effect of EPD processing time on capacitance

From the CV curves the capacitance of the electrodes can be characterized by integrating the areas under the curves[15]. Capacitance values of approximately 0.1, 3.2, 5.2, and 7.4 mF/cm^2 were calculated for the various processing times of 0, 3, 5, and 8 minutes at 20mV/s. Figure 15 shows a plot of the relationship observed between EPD processing time and the capacitance of the SWCNT coated SST electrodes. From this plot there appears to be a linear relationship between the EPD processing time and the capacitance of the SWCNT electrode for processing times less than 10min.

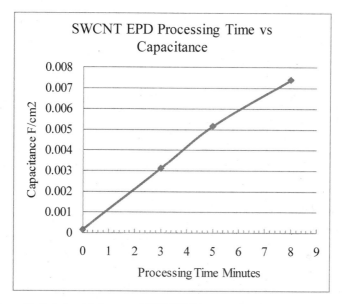

Figure 15: Relationship between SWCNT EPD processing time and capacitance

The addition of SWCNTs to SST by EPD greatly improves the capacitance of the bare SST electrodes. Comparing the SWCNT coated electrode which was processed for 8min to the bare SST electrode the capacitance of the SWCNT coated electrode was 45 times larger than the uncoated SST electrode. From the plot it can be hypothesized that further processing time could yield even greater increases in capacitance. Future work will investigate the effect of extended processing times >10 minutes to look for a limit to the linear relationship between capacitance and processing time as well as look at controlling EPD variables such as SWCNT solution concentration, EPD voltage, and EPD cell distance in order to produce SWCNT coated electrodes with higher capacitance in less processing time.

CONCLUSIONS

A process has been illustrated which is suitable for large scale high volume production of SWCNT coated supercapacitor electrodes from bulk SWCNT powders using EPD. The formation of uniform SWCNT networks on the surface of Ni and SST electrodes with nanopore diameters of approximately 20nm has been demonstrated. By maintaining the EPD processing voltage above 40V and using a processing time greater than 3min a repeatable process has been identified for coating SST and Ni electrodes with SWCNTs. The process has shown that an increase in the applied voltage and processing time results in an increase in SWCNT network coverage. The increase in processing time has been characterized at 40V and shows that increases in EPD processing time directly correspond to increases in electrode capacitance. In this work the addition of a SWCNT network to a SST electrode by EPD resulted in an increase in capacitance of 45 times the capacitance of the uncoated electrode.

ACKNOWLEDGMENTS

The authors would like to acknowledge the financial support of the Natural Sciences and Engineering Research Council of Canada and the materials support of Nano-C.

REFERENCES

[1] A.G. Pandolfo, A.F. Hollenkamp, Carbon properties and their role in supercapacitors, Journal of Power Sources 157, 11-27, (2006)

[2] A.S. Aricò, P. Bruce, B. Scrosat, J. Tarascon, W. van Schalkwijk, Nanostructured materials for advanced energy conversion and storage device, Nature Materials 4, 366 - 377 (2005)

[3] B.E. Conway, Electrochemical Super-capacitors: Scientific, Fundamentals, and Technology Applications, Kluwer, New York (1999)

[4] J. Chmiola, G. Yushin, Y. Gogotsi, C. Portet, P.Simon, P.L. Taberna, Anomalous Increase in Carbon Capacitance at Pore Sizes Less than 1 Nanometer, Science, vol. 313, 1760-3, (2006)

[5] S. Patrice, A. Burke, Nanostructured carbons: double-layer capacitance and more, Electrochemical Society Interface, vol.17, no.1, .38-43, (2008)

[6] J. Li, Q.M. Yang, I. Zhitomirsky, Nickel foam-based manganese dioxide-carbon nanotube composit electrodes for electrochemical supercapacitors, Journal or Power Sources, 185, 1569-1574, (2009)

[7] A.R. Boccaccini, I. Zhitomirsky, Applications of electrophoretic and electrolytic deposition techniques in ceramics processing, Solid State and Material Science, 6, 251-260 (2002)

[8] G. Girishkumar, M, Rettker, R. Underhile, D. Binz, K. Vinodgopal, P. McGinn, P. Kamat, Single-Walled Carbon Nanotube-Based Proton Exchange Membrane Assembly for Hydrogen Fuel Cells, Langmuir, 21, 8487-8494, (2005)

[9] M. Kaempgen, C. K. Chan, J. Ma, Y. Cui, G. Gruner, Printable Thin Film Supercapacitors Using Single-Walled Carbon Nanotubes, Nano Letters, 9 (5), 1872-1876, (2009)

[10] T. Kitano, Y. Maeda, T. Akasaka, Preparation of transparent and conductive thin films of carbon nanotubes using a spreading/coating technique, Carbon, 47, 3559-3565, (2009)

[11] B.A. Kakade, V. K. Pillai, An efficient route towards the covalent functionalization of single walled carbon nanotubes, Applied Surface Science, 254, 4936-4943, (2008)

[12] M. Shaffer, X. Fan, A.H. Windle, Dispersion and Packing of Carbon Nanotubes, Carbon, 36, 1603–1612, (1998)

[13] J. Cho, K. Konopka, K. Rozniatowski, E. Garcia-Lecina, M. Shaffer, A.R. Boccaccini, Characterisation of carbon nanotube films deposited by electrophoretic deposition, Carbon, 47, 58-67,(2009)

[14] A. Ma, K. Yearsley, M. MacKley, F. Chinesta, A review of the microstructure and rheology of carbon nanotube suspensions, Journal of Nanoengineering and Nanosystems, 222, 71-94, (2008)

[15] C. Portet, J. Chmiola, Y. Gogotsi, S. Park, K. Lian, Electrochemical characterizations of carbon nanomaterials by the cavity microelectrode technique, Electrochimica Acta, 53, 7675-7680, (2008)

THE MECHANOCHEMICAL FORMATION OF FUNCTIONALIZED SEMICONDUCTOR NANOPARTICLES FOR BIOLOGICAL, ELECTRONIC AND SUPERHYDROPHOBIC SURFACE APPLICATIONS

Steffen Hallmann*, Mark J. Fink[§] and Brian S. Mitchell*[#]

* Department of Chemical and Biomolecular Engineering, Tulane University, 300 Lindy Boggs, New Orleans, LA, 70118 (USA),
E-mail: brian@tulane.edu
[§] Department of Chemistry, Tulane University, 2015 Percival Stern Hall, 6283 St. Charles Ave, New Orleans, LA, 70118 (USA),
E-mail: fink@tulane.edu

[#] Corresponding author: Department of Chemical and Biomolecular Engineering, Tulane University, 300 Lindy Boggs, New Orleans, LA, 70118 (USA), E-mail: brian@tulane.edu, Phone: 504-865-8257, Fax: 504-865-6744

ABSTRACT

A facile method for the fabrication of functionalized semiconductor nanoparticles is described. The mechanochemical method involves the simultaneous top-down formation of nanoparticles using high energy ball milling and reaction with a liquid medium, such as alkynes and alkenes, to functionalize the nanoparticle surfaces as they are formed. As the silicon fractures during the mechanical attrition, the formation of reactive surface species leads to surface functionalization due to addition reactions with the surrounding organic medium and the establishment of stable Si-C bonds. As the particles further fracture into the nano-regime and become sufficiently functionalized with organic molecules, they become soluble in the parent liquor. This process can be adapted to form water soluble functionalized silicon nanoparticles. Potential applications as biomarkers and solar energy collection media are discussed. Finally, the use of these materials for the production of superhydrophobic films and surfaces is described.

INTRODUCTION

Funtionalized silicon-based materials are used in a wide range of applications due to their abundance and unique physical properties. In particular, silicon nanoparticles are attracting growing interest due to their use in optoelectronic devices, [1-3] solar cells,[4] nanoparticle lasers,[5,6] and as fluorescent biomarkers[7-9] with low cytotoxicty. Quantum confinement of electrons in silicon nanoparticles with a crystallite radius less than the Bohr exciton radius (5 nm for silicon) results in size dependent bandgaps[10, 11] and thus a Stokes shift, which causes photoluminescence with high quantum yields.[12] Functionalized nonporous alkyl functionalized silicon particles of sizes bigger than 100 nm exhibit properties similar to bulk silicon. However, in depth studies of these alkyl functionalized micrometer sized silicon particles are lacking due to the difficulties associated with monocrystalline formation and simultaneous surface functionalization. Recently the formation of alkyl functionalized, monocrystalline silicon particles and their application as surface coatings with superhydrophobic properties was described,[13] which leads to possible applications for solar cells and nontoxic, antioxidation coatings using mechanochemical synthesis. Mechanochemical synthesis via High Energy Ball Milling (HEBM) has been shown to be a facile and efficient method for the formation of alkyl passivated silicon nanoparticles and microparticles.[13-15] HEBM of silicon wafers carried out in an reactive environment like unsaturated organic compounds results in the formation of stable Si-C bonds and thus functionalization with alkyl chains. In the present study the alkyl passivation of the silicon surface is obtained through milling in a reactive liquid medium like alkynes or alkenes under inert atmosphere in a stainless steel milling vial. The reaction of a terminal triple or double bond with the reactive Si=Si and silicon surface radicals results in the formation of a covalent Si-C bond, which prevents further oxidation of the silicon surface.[16] This process forms alkyl functionalized silicon particles ranging in diameter from a few nanometers to several micrometers. The alkyl functionalized silicon nanoparticles are soluble in organic solvents including the organic liquid used during the mechanochemical synthesis. This phenomenon is used to separate the soluble nanoparticles from the micrometer sized particles by centrifugation (Figure 1). After the separation, the clear, usually light yellow, supernatant contains the nanoparticles and shows a strong blue luminescence,[14,15] whereas the sediment consists of the micrometer sized alkyl functionalized particles which can be used to form coatings with superhydrophobic properties.

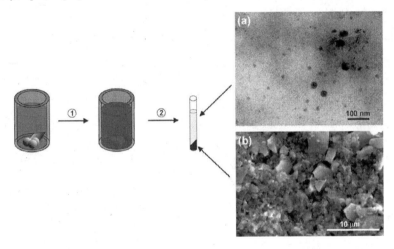

Figure 1: Shematic illustration of HEBM process. Step 1: A milling vial is filled with silicon wafer, three milling balls and the reactive organic medium; and milled for 12 - 24 hrs. Step 2: The resulting slurry is centrifuged, where the (a) supernatant contains alkyl functionalized silicon nanoparticles and (b) the sediment contains alkyl functionalized micrometer sized silicon particles (b).

Superhydrophobic silicon surfaces might have applicability not only in photovoltaics, but also in a wide variety of other applications, such as anti-fogging and anti-oxidation coatings.[17] In contrast to most silicon based procedures including lithographic pattering,[18-21] vapor deposition,[22,23] utilization of templates,[24] chemical deposition[25-27] self-assembled monolayers,[28] layer-by-layer (LBL),[29,30] sol-gel methods with phase separation,[31,32] plasma etching,[33,34] and bottom-up approaches[35] HEBM can be scaled up to industrial amounts.

In this communication we will show that HEBM can be used to form a variety of alkyl surface functionalizations, which can alter the surface characteristics of the silicon particles.

Furthermore, strategies are presented to obtain water soluble silicon nanoparticles via this method and how wettability of the produced microparticles can can be influenced by milling time, surface functionalization and deposition solvent.

EXPERIMENTAL

Nanoparticle Formation

All chemicals were purchased from Sigma-Aldrich or TCI America, and employed without further purification unless otherwise specified. Hexyne (97% assay) was distilled to remove halogen contaminants. To produce the chloroalkyl group functionalized silicon nanoparticles 0.75 g to 2.5 g of crystalline silicon (as obtained from Silrac, undoped, mirror finish, orientation [111]) was milled in various concentrations of 6-chloro-1-hexyne in hexyne or hexene with a total liquid volume of 25 ml. To form octyl and propyl alcohol capped silicon nanoparticles 0.75 g of crystalline silicon (as obtained from Silrac, undoped, mirror finish, orientation [111]) was milled in 25 ml octyne or ally lalcohol. The stainless steel milling vial was loaded with the appropriate chemical components and three stainless steel milling balls under inert nitrogen atmosphere to prevent oxidative side reactions during the milling process. The milling was performed for 12 hours in a SPEX 8000D high energy ball mill in a cold room at 2 °C. After the milling was completed the vial contents were transferred to a plastic centrifugation tube and centrifuged for 30 min at 511 G. The clear supernatant was transferred to a glass vial and the solvent was removed under reduced pressure in a vacuum oven at 25 °C. The oily residue containing the nanoparticles was then dissolved in organic solvents such as dichloromethane. The nanoparticle yield of this process varies between 60 to 120 mg for a 0.75 g batch, depending on the chemical composition of the organic liquids and thus reactivity. This equals a nanoparticle yield of 8 % - 16 % to the silicon precursor mass.

Micelle Encapsulation

Typically, 5.5 μmol of 60 % n-poly (ethyleneglycol) phosphatidylethanol-amine (PEG-PE) and 40 % phosphatidylcholine (PC) was dispersed with octyl functionalized silicon nanoparticles[14] in 5 ml chloroform and stirred for one hour. The solvent was evaporated in a vacuum oven at room temperature. The residue was solved in deionized water and subjected to ultrasonication for one hour using a bath sonicator. The resulting dispersion was passed through a 0.2 μm membrane filter. SDS encapsulation was carried out using the same procedure but with 5 mg of SDS instead of phospholipids. The sodium dodecyl sulfate (SDS) encapsulated nanoparticles were used for gel electrophoresis using 2 % agarose gel and a voltage of 80 V.

Particle Coatings

The sediment containing the micrometer sized functionalized silicon particles recovered from centrifugation was stirred for 10 min before coating a desired surface (usually a 25 mm x 25 mm glass slide) through drop-wise deposition of 1 ml of the alkyl functionalized silicon particles in different solvents resulting in a mass of about 0.3 g of particles. A nanoparticle solution was produced from the remaining centrifugation supernatant according to previous published protocols.[13] For characterization, films from the micrometer sized particles as well as the nanoparticle solutions were deposited on 25 mm x 25 mm glass slides as described above. The coated surface was dried for 10 min in air at room temperature and then exposed to a reduced atmospheric pressure for 10 min. Weight controls of samples dried for longer times under reduced atmospheric pressure and elevated temperature revealed that after this procedure the coating is completely dry and is approximately 100 μm thick.

Characterization
 Fourier Transformed Infrared (FT-IR) Spectroscopy was performed using a Thermo Nicolet NEXUS 670 FT-IR. The same unit, with an attached NXR FT-Raman Module, was additionally used for Raman Spectroscopy. FT-IR was achieved from deposit films of functionalized silicon nanoparticles and sediment on a KBr plate. Nuclear magnetic resonance spectroscopy was conducted in chloroform-d_l using a Bruker Avance 300 mHz high resolution NMR spectrometer. Transmission electron microscopy (TEM) images were taken with a JOEL 2011 TEM using accelerating voltage of 200 kV. Energy dispersive spectroscopic (EDS) data were obtained in the TEM using an Oxford Inca attachment with a 3 nm beam spot on a copper grid. The photoluminescence data from the nanoparticles in dichloromethane were obtained using a Varian Cary Eclipse spectrofluorimeter. UV-Vis absorbance characteristics in dichloromethane were acquired with a Cary 50 spectrophotometer. The static and dynamic contact angles of the sediment were measured with a Standard Goniometer (ramé-hart Model 250) with DROPimage Advanced v2.4. The volume increase/decrease used to measure the dynamic contact angles was 0.2 µl/s. For scanning electron microscopy (SEM) and energy disperse X-ray spectroscopy (EDS), the Hitachi S-3400N scanning electron microscope with INCAx-sight from Oxford Instruments was utilized.

RESULTS AND DISCUSSION
 In a previous publication it was shown that HEBM is a powerful tool for the formation of alkyl passivated, hydrophobic nanoparticles with a strong blue luminescence.[14] However, for some applications, in particular environmental and biological, the nanoparticles must be soluble in hydrophilic solvents such as water. The encapsulation of the hydrophobic nanoparticles in lipids and the milling in bifunctional organic molecules are two techniques to achieve this goal.

Lipid encapsulation
 Dubertert et al first showed the general use of lipids to encapsulate hydrophobic quantum dots for bioimaging using a mixture of n-poly (ethyleneglycol) phosphatidylethanol-amine (PEG-PE) and phosphatidylcholine (PC).[36] The same approach was used here for octyne capped silicon nanoparticles produced by HEBM. The hydrophobic tails encapsulate the silicon nanoparticles with the octanyl surface groups and thus change the solubility of the nanoparticles by exposing the hydrophilic components of the lipids. This process results in nanoparticle-containing micelles ranging from ~ 5 to ~50 nm, determined by TEM measurements (Figure 2a). The obtained EDS spectrum shows a high concentration of silicon in this sample and thus proves the successful encapsulation (Figure 2b). The EDS also shows that next to the elements carbon (~0.3 keV), sodium (~1.1 keV), oxygen (~0.55 keV) from the phospholipid solution, silicon (~1.8 keV) from the Si-NP and copper (~8.1 and ~9 keV) from the Cu-grid, elements like potassium (~3.4 keV) and chlorine (~2.65 keV) are detectable. These elements probably originate from the phospholipid solution, since they cannot be detected in any other samples without surfactant.
 The TEM of the PEG-PE and PC encapsulated particles shows a relatively broad size distribution, which might be due to the size variation of the nanoparticles in the precursor solution exclusively and/or due to the encapsulation of more than one nanoparticle in one micelle. To explore these possibilities the nanoparticles were encapsulated in a lipid which is able to facilitate a size separation. For this approach SDS was used, which establishes a negative charge on the nanoparticle surface and thus can be employed for size separation using electrophoretic methods. The TEM of these samples show the same size range as with the PEG-PE and PC system (Figure 2c). The EDS spectrum shows exclusively the anticipated elements

silicon, carbon, oxygen and cooper, but no potassium or chlorine since a higher grade lipid was available.

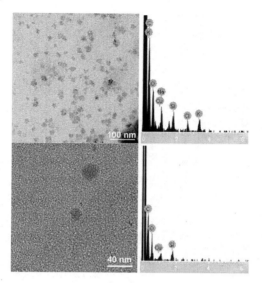

Figure 2: TEM and EDS images of (a) PEG lipid encapsulated silicon nanoparticles and (b) SDS encapsulated silicon nanoparticles.

The ability to size separate the encapsulated nanoparticles can be used to determine whether multiple or single nanoparticles are within the lipid micelle. The strong blue luminescence of silicon nanoparticles can be used to identify the nanoparticles after size separation. This strong blue photoluminescence is also preserved in the lipid encapsulated samples and seems to be uninfluenced by the choice of lipids (Figure 3a, b). In the case of single nanoparticle encapsulatuion one strong band of blue photoluminescence should appear after size separation, since the blue emitting nanoparticles are in a narrow size range. In the case of multiple nanoparticle encapsulation this band should be replaced by a long stretched band due to various sizes of the micelles, with very large micelles and their agglomerates separated from it at the beginning of the agarose gel. The second instance is indeed observed, when run in a 2% agarose gel at a voltage of 80 V (Figure 3 c). From this result we conclude that the wide distribution of observed sizes of lipid encapsulated nanoparticles in the TEM (Figure 2) is based on micelles containing multiple numbers and sizes of nanoparticles.

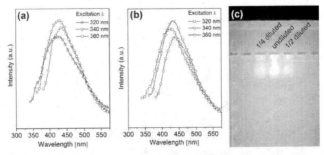

Figure 3: Photoluminescence spectra of (a) PEG lipid encapsulated silicon nanoparticles and (b) SDS encapsulated silicon nanoparticles. (c) UV-light image of size separated silicon nanoparticle SDS composite in agarose gel.

Mechanochemical synthesis of water soluable silicon nanoparticles

The lipid encapsulation to obtain water soluble nanoparticles utilizes octyl passivated silicon nanoparticles, produced by mechanochemical synthesis using the monofunctionalized hydrocarbon octyne. Another more direct approach is to use a bifunctionalized organic compound like ally lalcohol. Here the double bond of this organic liquid primarily reacts with the reactive silicon surface and thus results in propyl alcohol functionalized silicon nanoparticles. In a previous communication it was shown that double bonds and alcohol groups react with the newly formed silicon surface,[15] however the difference in the reactivity favors the addition of the double bond to the reactive silicon sites (Figure 4c). The dominant reaction with the double bond of the ally lalcohol, which is present in excess during the mechanochemical synthesis is indicated by the lack of absorbance bands for a double bond at 1650 cm^{-1} and 3040 cm^{-1} (Figure 2b). This is corroborated by strong alcohol specific absorbance bands, in particular broad υ_s(OH) stretching vibration ~3330 cm^{-1} and the primary alcohol associated υ_s(C-O) stretching vibration at ~1370cm^{-1} and the δ_s(C-O) deformation vibration at ~1030 cm^{-1}. The successful reaction with the double bond is further indicated by Si-C bands at 1251 cm^{-1} and 849 cm^{-1}. The absorbance peak at 1733 cm^{-1} is attributed to process specific impurities*.

---

* Plasticizer can leach out from O-Rings and centrifugation tubes during the process of High Energy Ball Milling . Results will be published in a separate paper.

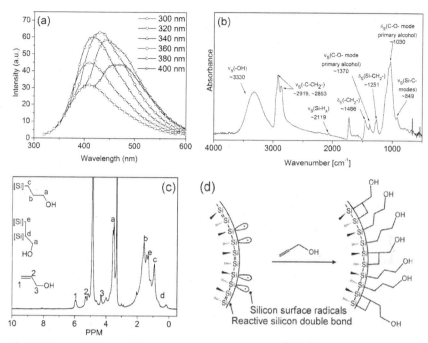

Figure 4: (a) Photoluminescence spectrum, (b) FTIR spectrum, (c) ^{1}H NMR spectrum of propyl alcohol functionalized silicon nanoparticles. (d) schematic illustration of the allyl alcohol grafting during the mechanochemical process.

The ^{1}H NMR spectrum from the allyl alcohol functionalized nanoparticles shows strong peaks for a primary alcohol group at 3.6 ppm and silicon bound alkyl groups at 1.0 ppm – 2.0 ppm (Figure 4c). This NMR also shows residual ally lalcohol indicated by the numbers one through three, which is likely due to nanoparticle adsorbed trace concentrations. The FTIR spectrum and the ^{1}H NMR spectrum prove the dominant reaction of the double bond of the ally lalcohol with the silicon, a fact further supported by the excellent solubility of the obtained nanoparticles in polar solvent like methanol and Dimethylformamide. The nanoparticles show a strong blue luminescence under UV radiation and the characteristic bathochromic shift of the emission wavelength with increasing excitation wavelength (Figure 6), typical of a polydisperse sample of nanoparticles.[37] The origin of the photoluminescence typically arises from particle size dependent core states but can be also influenced by surface states, oxidation and crystal defects of the silicon nanoparticles.[38-41] The photoluminescence spectrum is in good agreement with previous published photoluminescence spectra for alkyl functionalized silicon nanoparticles, which show a strong blue luminescence for small silicon nanoparticles of sizes up to 5 nm.[42-45]

The milling of silicon in ally lalcohol is one mechanism to produce nanoparticles with alkyl linked functional groups. Other organic milling compounds include chloroalkenes and chloroalkynes like allyl chloride and 6-chloro-1-hexyne. These compounds yield silicon nanoparticles with a highly reactive terminal chloro group, which can be used for a variety of secondary reactions to graft sugar molecules, dendrite monomers or PEG compounds to the functionalized silicon nanoparticles; and to replace the chloro group with hydroxyl and amino groups by well known organic reaction.[46]

ALKYL FUNCTIONALIZED SILICON PARTICLES WITH SUPERHYDROPHOBIC PROPERTIES

HEBM has been shown to be an alternative method to produce micrometer sized particles, which can form films with superhydrophobic properties.[13] Superhydrophobic surfaces usually exhibit by a combination of micro- and nanometer scale structures, which are vital for the superhydrophobic properties.[47] The deposited particles films obtained from HEBM show these structural features through a wide size distribution from the nanometer to the micrometer regime of alkyl functionalized particles. The resulting films are stabilized by hydrophobic interactions between the alkyl chains on the surface of the particles[48-50] and thus can coat different substrates with various surface morphologies.

Previously we characterized alkyl functionalized silicon particles from HEBM obtained by milling between 5 min and 12 hrs. Here the decreasing particles size with increasing milling time resulted in an increasing contact angle of water due to a change in the ratio of nanometer and micrometer sized particles. Films formed from particles produced by milling times smaller than 4 hrs showed a wetting behavior related to the Wenzel mode,[51] whereas longer milling times produce particle films with wetting properties related to the Cassie Baxter mode.[52] This shift from films were water can access the cavities between the particles (Wenzel mode) to films where air is trapped in the cavities and water is unable to enter them (Cassie-Baxtor mode), is caused by a change in the surface morphology based on different distribution of particles in the nanometer and micrometer range. The wetting properties of these films are influenced by milling time, alkyl grafted polar groups, and deposition using solvent mixtures of various polarities.

The effect of increased milling times

The milling time greatly influences the particles size distribution of the alkyl functionalized silicon particles collected as sediment during centrifugation of the crude milling slurry. This variation of particle sizes changes the ratio of nanometer sized to micrometer sized particles in the sediment and thus the surface roughness of the films. The smaller particles can access the void volume between bigger particles (up to 50 μm) and prevent water from accessing the cavities between the particles. Previously it was shown that this effect causes a decreasing wettability up to a milling time of 12 hours, with a local minimum of the contact angle at 8 hours.[13] The ratio of small particles increases further with milling times exceeding 12 hrs and thus results in a continuous decrease of the surface roughness (Figure 5a). However the contact angle seems to reach a maximum at a milling time of 12 hours at around 150 ° and starts to decrease after this time (Figure 5b). This suggest that the surface roughness becomes too low to support a high contact angle, which is in agreement with superhydrophobic surfaces produced by different methods.[30-46]

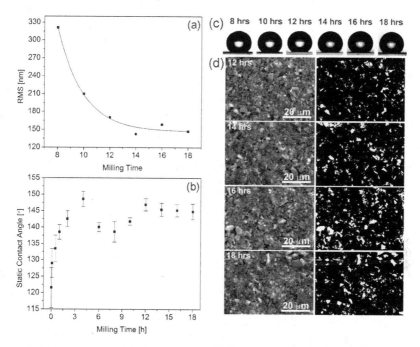

Figure 5: (a) AFM measured surface roughness, b) Static contact angle, (c) Drop images and (d) SEM images of films formed by particles synthesized by the indicated milling times.

This conclusion is supported by the calculation of the contact angle using the well known Cassie-Baxter equation.

$$\cos\theta^{CB} = f_1 \cdot \cos\theta^Y - f_2 \tag{1}$$

Here the fraction f_1 corresponds to the area of liquid in contact with the surface divided by the projected area and, f_2, to the area in contact with the trapped air beneath the liquid drop divided by the projected area as well.

To estimate the fraction f_1 a calculation based on ImageJ developed by the National Institute of Health was introduced by Y. Wang (Figure 5d).[53] The calculation was carried out for films obtained from milling times of 12, 14, 16 and 18 hours. For these times the values of f_1 are 7.2%, 10.1 %, 13.7 and 11.1 % at a constant threshold and indicate a increasing surface in contact with the droplet for smaller surface roughness values. By using equation 1 and the Young contact angle of 87 ° from a flat alkyl functionalized silicon nanoparticles surface, the Cassie Baxter contact angles of 148.8 ° (12 hrs), 143.1 ° (14 hrs), 136.6 ° (16 hrs) and 141.2 ° (18 hrs) were calculated. These values are in good agreement with the measured values in Figure 5 b. The results prove that the optimal size distribution of alkyl functionalized silicon particles is obtained after a milling time of 12 hrs. Films produced by particles from milling experiments with longer process times exhibit too low of a surface roughness to maintain a superhydrophobic contact angle.

The effect of alkyl linked chloro groups

The superhydrophobic properties of the produced films are greatly influenced by the surface state of the functionalized silicon particles. By introducing terminal chloro alkyl groups on the surface of the particles this surface state can be changed towards an increased polarity and thus a higher wettability. The terminal chloro alkyl groups can be grafted to the silicon particle surface by milling silicon wafer in a coreactant mixture of hexene and 6-chloro-1-hexyne. This process can be adapted to fine tune the surface coverage of the alkyl linked chloro groups and thus analyze the influence of a slowly increasing polarity on the resulting contact angle. The coverage of the terminal chloro alkyl groups of the resulting particles was analyzed by ^{1}H NMR to identify the final surface concentration of the alkyl linked chloro groups. Films produced from these particles show an decreasing static contact angle up to a chloro alkyl concentration of 17.5 %. At higher concentration the static contact angle stays constant at 139 ° (Figure 6a). However, the influence on the dynamic advancing and receding contact angle seems to follow a more linear trend (Figure 6b and 6c). This effect can be explained by a higher sensitivity of the dynamic contact angles to changes of the surface states. Especially the receding dynamic contact angle with a decrease of more than 50 % of the initial value is highly influenced by the hydrophobicity of the surface.

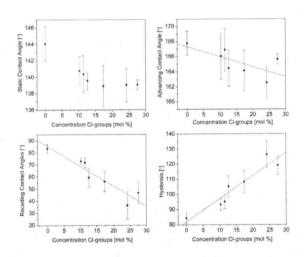

Figure 6: Static and dynamic contact angles measurements of films formed by silicon particles with various degrees of chloroalkyl coverage.

These results show that the static contact angle is mainly influence by the surface roughness (Figure 5) and less by the chemical composition of a hydrophobic surface. The dynamic contact angles in contrast are very sensitive to the surface polarity as well as the overall surface roughness.

The effect of the polarity of the deposition solvent
The superhydrophobic films are produced by deposition of the alkyl functionalized silicon particles in hydrophobic solvents like hexane (Figure 7a). The hydrophobic solvent mediates uniform distributions of the different sized particles since the hydrophobic attraction forces between the solvent and particles are equal. However, by increasing the polarity of the deposition solvent the attraction forces of the particles to each other can overcome solvation forces and cause the particles start to agglomerate (Figure 7b). [54]

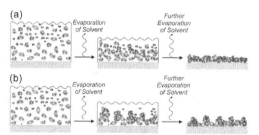

Figure 7: Shematic illustration of the deposition process of sfunctionalized silicon particles with (a) nonpolar solvents and (b) polar solvents.

This effect can be used to induce a change in the surface morphology during the deposition process, which can increase the superhydrophobic character of the films when deposited with solvents of higher polarity. In extreme cases the attraction forces of the alkyl functionalized particles to each other can exceed the solvation force to the extent that the films rupture during the deposition process (Figure 8).
To examine this solvent driven agglomeration an increasing concentration of ethanol in hexane was used as deposition solvent. The films prepared with various concentration of ethanol show an increase of the static contact angle from 145.5 ° to 155 ° (Figure 8 b).

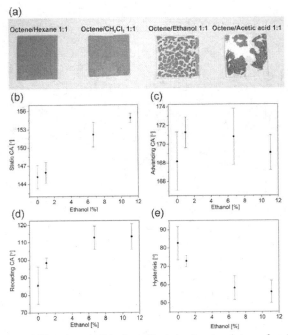

Figure 8: (a) Image of films deposited in the indicated solvents mixture of various polarities. (b) static and (c, d) dynamic contact angles as well as (e) hysteresis of silicon particles films deposited with different ethanol concentrations.

The dynamic advancing contact angle on the other hand changes only slightly, which can be explained by the high initial value (Figure 8 c). In contrast, the dynamic receding contact angle with a low initial value increases by about 35 % and reaches a maximum of around 115 ° at 8 % ethanol in an asymptotic fashion (Figure 8 d). The resulting hysteresis shows a similar behavior and corroborates the increased superhydrophobic character of films produced by deposition in solvents with increased polarity. Initial SEM and AFM images (not shown) indicate the formation of grapevine like structure around bigger particles illustrated by Figure 7. These structures could reduce the surface in contact with the water droplet, when arranged in a certain distance and thus increase the contact angles (Equation 1). However the differences in AFM and SEM images are minimal and thus difficult to detect. From the obtained result we therefore conclude that a higher polarity of the deposition solvents decreases the wettability by secondary surface rearrangements.

CONCLUSION
We have demonstrated that mechanochemical synthesis via high energy ball milling is a facile and fast method to simultaneous produce and functionalize silicon nanoparticles with strong photoluminescence and silicon particles for superhydrophobic application. Water solubility of these particles can be obtained by lipid encapsulation of alkyl passivated silicon

nanoparticles. The encapsulation with PEG lipids makes the nanoparticle usable for biological application. The encapsulation with SDS results in charged nanoparticles composites and thus can be used for electrophoretic purification techniques. The mechanochemical synthesis using high energy ball milling also produces high quantities of functionalized micrometer sized particles, which can be used to produce coatings with superhydrophobic properties. The wettability is greatly influenced by the size distribution of the functionalized silicon particles. After a milling time of twelve hours this distribution seems to be optimal and results in films with a maximum contact angle of around 150 °. Longer millings times decrease the surface roughness and thus decrease the measured contact angle. To further increase the contact angle the particles can be deposited in solvents of higher polarity, which mediates agglomeration processes and thus changes the surface morphology of the films favoring higher contact angles.

ACKNOWLEDGMENTS

This work is supported by the National Science Foundation (NSF grant CMMI- 0726943) and by the Tulane Institute for Macro-Molecular Engineering and Sciences (TIMES, NASA Grant NNX08AP04A).
References

REFERENCES
[1] S. Pillai, K. Catchpole, T. Trupke, G. Zhang, J. Zhao, M. Green. *Appl. Phys. Lett* **2006**, 88, 161102.
[2] L. Raniero, S. Zhang, H. Aqyas, I. Ferreira, R. Igreja, E. Fortunato, R. Martins, *Thin Solid Films* **2005**, 487, 170.
[3] O. Boyraz, B. Jalili, *Optics Express* **2005**, 13, 796.
[4] M. C. Beard, K. P. Knutsen, P. Yu, J. M. Luther, Q. Song, W. K. Metzger, R. J. Ellingson, A. J. Nozik, *Nano Lett.* **2007**, 7, 2506.
[5] A. Fotjik, J. Valenta, I. Peant, M. Kalal, P. Fiala, *J. Mater. Process Technol.* **2007**, 181, 88.
[6] G. Belomoin, J. Therrien, A. Smith, S. Rao, R. Twesten, S. Chiab, M. H. Nayfeh, L. Wagner, L. Mitas, *Mater. Res. Soc. Symp. Proc.* **2002**, 703, 475.
[7] L. Wang, V. Reipa, J. Blasic, *Bioconjug. Chem.* **2004**, 15, 409.
[8] D. R. Larson, W. R. Zipfel, R. M. Williams, S. W. Clark, M. P. Bruchez, F. W. Wise, W. W. Webb, *Science* **2003**, 300, 1434.
[9] M. Rosso-Vasic, E. Spruijt, Z. Popovi, K. Overgaag, B. van Lagen, B. Grandidier, D. Vanmaekelbergh, D. Dom•nguez-Gutiérrez, L. De Cola and H. Zuilhof, *J. Mater. Chem.*, **2009**, 19, 5926–5933.
[10] J. M. Buriak, *Chem. Commun. (Camb)* 1999, 105.
[11] J. P. Proot, C. Delrue, G. Allan, *Appl. Phys. Lett.* **1992**, 61, 1948.
[12] M. H. Nayfeh, E. V. Rogozhina, L. Mitas, Synthesis, *Funtional. Surf. Treatment Nanopart.* **2003**, 173.
[13] S. Hallmann, B. S. Mitchell and M. J. Fink, *Journal of Colloid and Interface Science*, **2010**, 348, 634.
[14] A. S. Heintz, M. J. Fink, B. S. Mitchell, *Adv. Material.* **2007**, 19, 3984.
[15] A. S. Heintz, M. J. Fink and B. S. Mitchell, *Appl. Organometal. Chem.* **2010**, 24, 236.
[16] D. R. Maurice and T. H. Courtney, *Metallurigal Transactions* A **1990**, 21, 289.
[17] A. Nakajima, K. Hashimoto, T. Watanabe, *Monatsh. Chem.* **2001**, 132, 3.
[18] R. Fürstner, W. Bartholtt, C. Neinhuius, P. Walzel, *Langmuir* **2005**, 21, 956

[19] L. Feng, S. H. Li, Y. S. Li, H. J. Li, L. J. Zhang, L. J. Zhai, Y. L. Song, B. Q. Liu, L. Jiang, D. B. Zhu, *Adv. Mater.* **2002**, 14, 1857.

[20] D. Öner, T. J. McCarthy, *Langmiur* **2000**, 16, 7777.

[21] J.-Y. Shiu, C.-W. Kuo, P. Chen, C.-Y. Mou, *Chem. Mater.* **2004**, 16, 561.

[22] F. Wang, S. Song, J. Zhang, *J. Chem. Commun.* **2009**, 4239.

[23] L. Huan, F. Lin, Z. Jin, J. Lei, Z. Daoben, *Langmuir* **2004**, 20,5659.

[24] Y. Wu, H. Sugimura, Y. Inoue, O. Takai, Chem. Vap. Deposition **2002**, 8, 47.

[25] L. Feng, S. Li, H. Li, J. Zhai, Y. Song, L. Jiang, D. Zhu, *Angew. Chem. Int. Ed.* **2003**, 41, 1221.

[26] F. Shi, Z. Wang, X. Zhang, *Adv. Mater.* **2005**, 8, 1005.

[27] M. F. Wang, N. Raghunathan, B. Ziaie, Langmuir 23 **2007**, 2300.

[28] N. Zhao, F. Shi, Z. Wang, X. Zhang, *Langmuir* **2005**, 21, 4713.

[29] Y. Song, R. P. Nair, M. Zou, Y. Wang, *Nano Res.* **2009**, 2, 143.

[30] T. Soeno, K. Inokuchi, S. Shiratori, *Trans. Mater. Res. Soc. Jpn.* **2003**, 28, 1207.

[31] H.-J. Tsai, Y.-L. Lee, *Langmuir* **2007**, 23, 12687.

[32] N. J. Shirtcliffe, G. McHale, M. I Newton, C. C. Perry, *Langmuir* **2003**, 19, 5626.

[33] Y. Xiu, D. W. Hess, C.P. Wong, *Journal of Adhesion Science and Technology* **2008**, 22, 1907.

[34] B. D. Washo, *Org. Coat. Appl. Polym. Sci. Proc.* **1982**, 47, 69.

[35] B. Balu, V. Breedveld, D. W. Hess, Langmuir **2008**, 24, 4785.

[36] B. Dubertret, P. Skourides, D. J. Norris, V. Noireaux, A. H. Brivanlou, A. Libchaber, *Science* **2002**, 298, 1759.

[37] M. Trznadel, A. Pron, M. Zagorska, R. Chrzaszcz, J. Pielichowski, *Macromolecules*, **1998**, 31, 5051.

[38] X. Wang, R. Q. Zhang and S. T. Lee, *Appl. Phys. Lett.* **2007**, 90, 123116.

[39] A. S. Heintz, M. J. Fink and B. S. Mitchell, *Appl. Organometal. Chem.* **2010**, 24, 236.

[40] L. Patrone, D. Nelson, V. I. Safarov, M. Sentis and W. Marine, *J. Appl. Phys.* **2000**, 87, 3829.

[41] Y. Q. Wang, Y. G. Wang, L. Cao, and Z. X. Cao, *Appl. Phys. Lett.* **2003**, 83, 3474.

[42] Y. He, Z.-H. Kang, Q.-S. Li, C. H. A. Tsang, C.-H. Fan, and S.-T. Lee, *Angew. Chemie.* **2009**, 121, 134.

[43] Z. Kang, Y. Liu, C. H. Tsang, D. D. D. Ma, X. Fan, N.-B. S.-T. Wong, Lee, *Adv. Mater.* **2009**, 21, 661.

[44] A.-W. Lin and D.-H. Chen, *Small* **2009**, 5 (1), 72.

[45] N. Shirahata, M. R. Linford, S. Furumi, L. Pei, Y. Sakka, R. J. Gates and M. C. Aplund, *Chem. Commun,* **2009**, 4684.

[46] S. Hallmann, M. J. Fink and B. S. Mitchell, to be submitted.

[47] X. J. Feng, L. Jiang, *Adv. Mater.* **2006**, 18, 3063.

[48] H. Li, X. Wang, Y. Song, Y. Liu, Q. Li, L. Jiang, D. Zhu, *Angew. Chem., Int. Ed.* **2001**, 40, 1743.

[49] S. Marre, F. Cansell, C. Aymonier, *Langmuir* **2008**, 24, 252.

[50] K. P. Hapgood, B. Khanmohammadi, *Powder Technology* **2009**, 189, 253

[51] L. Forny, I. Pezron, K. Saleh, P. Guigon, L. Komunjer, *Powder Technology* **2007**, 171, 15.

[52] R. N. Wenzel, *Ind. Eng. Chem.* **1936**, 28, 988.

[53] A. B. D. Cassie, S. Baxter, *Trans. Faraday Soc.* **1944**, 40, 546.

[54] I. K. Snook, W. van Megen, *Phys. Lett A.* **1979**, 74, 332.

SYNTHESIS OF ZNO NANOSTRUCTURES AND THEIR INFLUENCE ON PHOTOELECTROCHEMICAL RESPONSE FOR SOLAR DRIVEN WATER SPLITTING TO PRODUCE HYDROGEN

Sudhakar Shet,[1, 2] Heli Wang,[1] Todd Deutsch,[1] Nuggehalli Ravindra,[2] Yanfa Yan,[1] John Turner,[1] and Mowafak Al-Jassim[1]

[1] National Renewable Energy Laboratory, Golden, CO, USA 80401
[2] New Jersey Institute of Technology, Newark, NJ 07102

ABSTRACT

We have investigated the structural properties, optical absorption, and PEC responses for ZnO nanostructures. All the films were deposited on fluorine doped tin oxide coated glass. We found that the presence of N_2 in the growth ambient help to promote the formation of aligned nanorods at high substrate temperature of 500°C, resulting in the significantly enhanced PEC response, compared to ZnO(Ar) films deposited in pure Ar gas ambient. A porous ZnO nanocoral structure synthesized using two step process provides a large surface area, superior light trapping, and an excellent pathway for carrier transport, resulted in a significantly increased PEC response compared to the compact ZnO thin films. Our results suggest that synthesis method can be used to produce desired properties of thin films and nanostructures may help to improve PEC performance.

INTRODUCTION

Photoelectrochemical (PEC) systems based on transition metal oxides, such as TiO_2, ZnO, and WO_3, have received extensive attention since the discovery of photoinduced decomposition of water on TiO_2 electrodes.[1-9] ZnO has similar bandgap (~3.3 eV) and band-edge positions as compared to TiO_2. ZnO has a direct bandgap and higher electron mobility than TiO_2.[8] Thus ZnO could also be a potential candidate for PEC splitting of water for H_2 production.[9]

To improve PEC performance, a photoelectrode should have a high contact area with the electrolyte to provide more interfacial reaction sites. Therefore, the morphological features of the thin films, such as grain size, grain shape, and surface area would have profound influence on the performance of thin film electrodes for PEC applications. It has been expected that electrodes with nanostructures would exhibit improved PEC performance as compared to those without nanostructures. Recently, ZnO electrodes with different nanostructures have been studied because of their potential application in optoelectronic nanoscale devices. It has been reported that ZnO nanorods were synthesized using catalyst, and recently, catalyst-free ZnO nanorod have been synthesized by various chemical and physical techniques such as metal-organic vapor-phase epitaxy, plasma-enhanced chemical vapor deposition and pulsed laser deposition.[10-13] So far, RF sputtering is much less considered than other methods for the growth of ZnO nanostructures and their influence on photoelectrochemical response. Detailed examination on this method for ZnO nanostructuresd growth and its effect on PEC response is needed.

In this paper, we report on synthesis technique to produce desired properties of ZnO thin films with nanostructures and their effect on PEC performance. We found that the presence of N_2 in the growth ambient help to promote the formation of aligned nanorods at high substrate temperature of 500°C, resulting in the significantly enhanced PEC response, compared to ZnO(Ar) films deposited in pure Ar gas ambient. A porous ZnO nanocoral structure synthesized using two step process provides a large surface area, superior light trapping, and an excellent pathway for carrier transport, resulted in a significantly increased PEC response compared to the compact ZnO thin films.

EXPERIMENTAL

For ZnO nanorods structures, two sets of samples are deposited. One set of samples was deposited in pure Ar gas ambient and is referred as ZnO(Ar). The second set of samples was deposited in mixed Ar and N_2 ambient. This set of samples is referred to as ZnO(Ar/N_2). ZnO targets were used in these depositions. Transparent conducting FTO (20–23 $\Omega/\square$)-coated glass was used as the substrate to allow PEC measurements. Prior to deposition, the substrates were ultrasonically cleaned by an acetone-methanol-deionized (DI) water sequence. The distance between the ZnO target and substrate was about 10 cm, and the substrates were rotated 30 rpm to enhance deposition uniformity. The base pressure was below 1×10^{-6} torr and the working pressure was 5×10^{-3} torr. The chamber ambient was either pure Ar or the mixed Ar and N_2 gas flow rate. Prior to sputtering, a pre-sputtering cleaning was performed for 20 min to eliminate possible contaminants from the target. Sputtering was then conducted at RF power of 300W for ZnO(Ar), and ZnO(Ar/N_2) samples. All the deposited samples were controlled to have similar film thickness of 1 ± 0.05 μm as measured by stylus profilometry.

ZnO nanocoral structures were synthesized by a two-step process: the deposition of Zn metal thin films followed by thermal oxidation at 500 °C in a quartz tube furnace with flowing O_2. Zn metal films were grown by RF magnetron sputtering. F-doped SnO_2 (FTO, 8-10 $\Omega/\square$)-coated transparent glasses (TEC-8, Hartford Glass Co.) were used as substrates for PEC applications. The substrates were rotated during deposition to achieve better film uniformity. The distance between the metallic Zn target (3-inch diameter) and the substrate was 8 cm. The base pressure was below 2×10^{-6} Torr, and the working pressure was 1.1×10^{-2} Torr under Ar ambient. A pre-sputtering process was performed for 10 min to eliminate any contaminants from the target. Sputtering was then conducted at room temperature at RF powers of 35, 50, 100, 150, and 200 W. All samples have a similar film thickness of about 1.4 μm as measured by stylus profilometry. After deposition, the Zn films were vacuum-sealed and kept in the N_2 desiccators to eliminate surface oxidization. ZnO nanostructures were formed by post-annealing the Zn films at 500 °C for 8 hr with the O_2 flowing (40 SCCM) in a three-zone quartz tube furnace.

The structural and crystallinity characterizations were performed by X-ray diffraction (XRD) measurements, using an X-ray diffractometer (XGEN-4000, SCINTAG Inc.), operated with a Cu Kα radiation source at 45 kV and 37 mA. The N concentration in the ZnO(Ar:N_2) films was evaluated by X-ray photoelectron spectroscopy (XPS). Monochromatic Al K_α radiation was used for all data sets, and the analyzer was set to 59 eV pass energy. Argon ion sputtering (3 keV, 0.8 μAmm^{-2}, 120 s) was used to clean samples prior to analysis. The surface morphology was examined by atomic force microscopy (AFM) conducted in the tapping mode with a silicon tip, and field emission scanning electron microscopy (FE-SEM) with a FEI Nova 200 SEM and transmission electron microscopy with a FEI Tecnai F20UT TEM.. The UV-Vis absorption spectra of the samples were measured by an n&k analyzer 1280 (n&k Technology, Inc.) to investigate the optical properties.

PEC measurements were performed in a three-electrode cell with a flat quartz-glass window to facilitate illumination to the photoelectrode surface.[14-30] The sputter-deposited films were used as the working electrodes. Pt plate and an Ag/AgCl electrode were used as counter and reference electrodes, respectively. A 0.5-M Na_2SO_4 mild aqueous solution was used as the electrolyte for the stability of the ZnO.[31] Photoelectrochemical response was measured using a fiber optic illuminator (150 W tungsten-halogen lamps) with a UV/IR filter. Light intensity was measured by a photodiode power meter, in which total light intensity with the UV/IR filter was fixed to 125 mW/cm^2.[14-31]

Because our films were deposited on conducting substrates, measurements of electrical property by the Hall Effect were not possible. Instead, the electrical properties were measured by Mott-Schottky plots, which were obtained by AC impedance measurements. AC impedance measurements were carried out with a Solartron 1255 frequency response analyzer using the above three-electrode cells. Measurements were performed under dark conditions with an AC amplitude of 10 mV and frequency of 5000 Hz were used for the measurements taken under dark condition and the AC impedances were measured in the potential range of -0.7 V to 1.25 V (vs. Ag/AgCl reference).[32-34] The series capacitor-resistor circuit model was used for Mott-Schottky plots.[35-36]

RESULTS AND DISCUSSION

We first see how the presence of N_2 in the ambient can promote the formation of aligned nanorods in ZnO thin films where the substrate temperature is higher than 300°C. Figures 1(a) and 1(b) show XRD curves for the ZnO(Ar) and ZnO(Ar/N$_2$) films deposited at different substrate temperatures. Dotted line indicates substrate peaks. The crystallinity of ZnO films increases gradually with the increase of substrate temperatures in both cases. However, with the increase of substrate temperature to 500°C, the (0002) peak of the ZnO(Ar/N$_2$) film was enhanced greatly, as shown in Fig. 1(b). The measured full-width at half-maximum (FWHM) values of (0002) peaks for ZnO(Ar) are approximately 0.21, 0.20, 0.19, and 0.19 for the films deposited at 200, 300, 400, and 500°C. The measured full-width at half-maximum (FWHM) values of (0002) peaks for ZnO(Ar/N$_2$) are approximately 0.25, 0.24, 0.22, 0.15, and 0.14 for the films deposited at 100, 200, 300, 400, and 500°C. The N concentrations (at%) for the ZnO(Ar/N$_2$) films measured by XPS are approximately, 0.32, 0.24 for the films deposited at 100 and 200°C and no detectable N concentration found for the films deposited at 300 or above temperature. With the increase of substrate temperature, the N concentration decreased rapidly and disappeared at temperatures above 300°C. The FWHM of ZnO(Ar) decreased slightly with the increase of substrate temperature. ZnO(Ar/N$_2$) films grown at temperatures below 200°C exhibited random orientation and larger FWHM values than the ZnO(Ar) films grown at the same temperatures. This is because the high concentration of N is incorporated in ZnO(Ar/N$_2$) films at these temperatures.

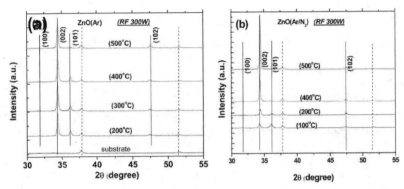

Figure 1. X-ray diffraction curves for (a) ZnO(Ar) films and (b) ZnO(Ar/N₂) films.

It is known from recent reports that incorporated N atoms can deteriorate the crystal structure and modify the growth mode.[15, 31-35] However, the FWHM values of ZnO(Ar/N₂) films decrease rapidly when the substrate temperatures are above 300°C because no significant N can be incorporated at these temperatures. At substrate temperatures above 300°C, ZnO(Ar/N₂) films exhibit much smaller FWHM values than the ZnO(Ar) films. The rapid growth of the FWHM values indicates either increased crystallinity or formation of nanorods or nanowires along the c-axis.

AFM images reveal that the significantly increased (0002) peak in the XRD curve obtained in ZnO(Ar/N₂) at 500°C is largely due to the formation of aligned nanorods along the c-axis. Figure 2 shows AFM surface morphology (5×5 μm²) of the ZnO(Ar/N₂) films deposited at the substrate temperatures of 100°, 200°, 400°, and 500°C (Figs. 2(a)–2(d), respectively), and the ZnO(Ar) film deposited at 500°C (Fig. 2(e)). It shows clearly that the ZnO(Ar/N₂) film deposited at 100°C has a random orientation. As substrate temperature increases, aligned nanorods along the c-axis are favored to form. At 500°C, the ZnO(Ar/N₂) film reveals the growth of hexagonal-like nanorods. However, the ZnO(Ar) film deposited at the same temperature is polycrystalline (Fig. 2(e)). It should be noted that at 500°C, the diameters of the nanorods are smaller than that of the grains in polycrystalline ZnO film. The smaller FWHM value for the ZnO(Ar/N₂) film is attributed to the nanorod feature.[36-38]

Figures 3(a) and 3(b) show FE-SEM top-views of the ZnO(Ar) and ZnO(Ar/N₂) films, respectively, deposited at a substrate temperature of 500°C. It clearly shows that the nanorod structure was not present in the ZnO(Ar) film, whereas the ZnO(Ar/N₂) films at 500°C exhibited vertically aligned, single-crystal hexagonal-like nanorods with flat (0002) surfaces. No metal clusters were found at the end of the nanorods, indicating that the growth mechanism is not the catalyst-assisted vapor-liquid-solid (VLS) growth.[37-39]

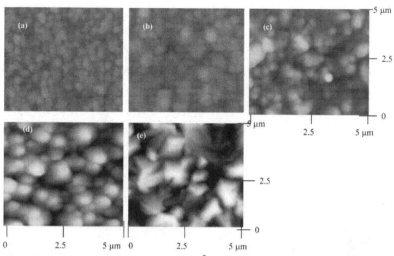

Figure 2. AFM surface morphology (5×5 μm^2) of (a-d) the ZnO(Ar/N$_2$)(300W) films deposited at the substrate temperatures of 100, 200, 400, and 500°C, and (e) ZnO(Ar)(300W) films deposited at500°C, respectively.

Figure 3. FE-SEM top-views of the (a) ZnO(Ar) film and (b) ZnO(Ar/N$_2$) nanorod film, respectively, deposited at 500°C.

Recently, catalyst-free ZnO nanorods/nanowires have been synthesized by various chemical and physical techniques such as plasma-enhanced chemical vapor deposition, metal-organic vapor-phase epitaxy, and pulsed laser deposition.[37-40] The nanorod structures provide high surface areas and superior carrier transport (or conductivity) along the c-axis, which may lead to increased interfacial reaction sites and reduced recombination rate.[10,41] Therefore, the aligned nanorod films deposited at 500°C should lead to enhanced PEC response.

The PEC response for the ZnO films deposited in different ambient was also investigated. Figures 4(a) and 4(b) show photocurrent-voltage curves of the ZnO(Ar/N$_2$) and ZnO(Ar) films deposited at 500°C, respectively, under continuous illumination (red curve), dark condition (black curve), with an UV/IR filter. Both ZnO films show very small dark currents up to a potential of 1.4 V. The ZnO(Ar/N$_2$) nanorod film deposited at 500°C exhibited much higher photocurrents than the ZnO(Ar) film deposited at the same substrate temperature.

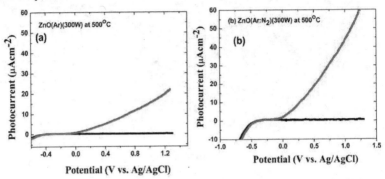

Figure 4. Photocurrent-voltage curves of (a) ZnO(Ar) films and (b) ZnO(Ar/N$_2$) nanorod films, deposited at 500°C under (red curve) continuous illumination, (black curve) dark condition, with an UV/IR filter.

The ZnO(Ar/N$_2$) film deposited at 500°C exhibits the best photoelectrochemical response more than two times higher than the ZnO(Ar) films deposited at the same temperature. The enhancement can be attributed to the aligned nanorod structure along the c-axis and additional light absorption in the long-wavelength regions. The electron-hole pairs are generated by the absorption of photons with energies larger than the bandgap and separated by the electric field of the depletion region. In the case of n-type semiconductors, the excited electrons move through bulk region to the counter electrode where water reduction occurs. The generated holes move towards semiconductor/electrolyte interface where water oxidation takes place. The aligned nanorod structures along the c-axis provide high surface area and superior carrier transport along the c-axis, leading to the increased interfacial reaction sites and reduced recombination rate between the electrons and holes. As a result, the PEC performance of the nanorod structure is greatly enhanced.

We now discuss about the ZnO nanocoral structure. Figure 5 shows the SEM images taken from the annealed samples of 35 W-ZnO, 50 W-ZnO, 100 W-ZnO, and 150W-ZnO, respectively. Clearly the 50 W-ZnO (Fig. 5b) and the 100 W-ZnO (Fig. 5c) samples exhibit a nanocoral structure. The average size of the nanocorals for the 50 W-ZnO is significantly smaller than that for the 100 W-ZnO samples, indicating that the size of the nanocorals can be tuned by the RF power. The 35 W-ZnO consists of nanoparticles, Fig. 5a. The 150 W-ZnO and the 200 W-ZnO (SEM image not shown here) are just nanocrystalline ZnO films, Fig. 5d. We note that these nanoparticle and nanocrystalline films are not as compact as films directly deposited as ZnO. The SEM images reveal two unique features for the nanocoral structures. First, because the nanocorals are composed of nanosheets, the structure is very porous, producing a large surface area. Second, the nanosheets grow together smoothly, providing an excellent electrical pathway for carrier collection. These two unique features are very favorable for PEC applications. [9, 42-44]

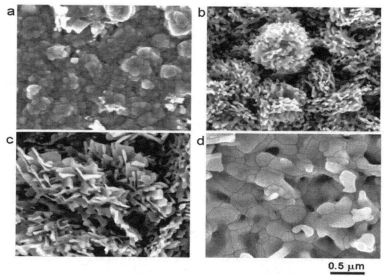

Figure 5. Microstructures of thermal oxidized ZnO films. **a, b, c, and d,** SEM images of 35 W-, 50 W-, 100 W-, and 150 W-ZnO films.

The above SEM images show that both 50 W-ZnO and 100 W-ZnO samples have nanocoral structures. These samples should exhibit enhanced PEC performance as compared to the normal compact ZnO thin films. The sample 50 W-ZnO has smaller nanosheets, larger surface area, and better crystallinity than the sample 100 W-ZnO. Thus, the 50 W-ZnO sample is expected to exhibit better PEC performance than 100 W-ZnO. These expectations have been confirmed by PEC response tests on various ZnO film morphologies. Figures 6a-6c show the photocurrent-voltage curves for the 50 W-, 100 W-, and 150 W- ZnO films, respectively, under chopped light illumination with an UV/IR filter. A 0.5-M Na_2SO_4 aqueous solution with a pH of 6.8 was used as the electrolyte. All of the samples showed very small dark currents up to a potential of 1.3 V, indicating that the photocurrents under light-on conditions are generated only by absorbed photons without a dark-current contribution. To compare the PEC responses, the photocurrents for all samples at 1.2 V were plotted, Fig. 6d. For comparison, the PEC responses of compact ZnO films reported elsewhere[14,16] are also plotted in the figure. Because all the films have similar thicknesses, the photocurrents can be compared. Clearly, the nanoparticle film (35 W-ZnO) and the nanocoral films (50 W-ZnO and 100 W-ZnO) exhibited higher PEC responses than compact ZnO films. Among them, the 50 W-ZnO nanocoral film exhibited the best PEC response and its photocurrent at 1.2 V is ten times higher than that of the compact ZnO films. On the other hand, the 150 W- and

200 W- ZnO films showed much lower PEC responses than the ZnO coral nanostructures.

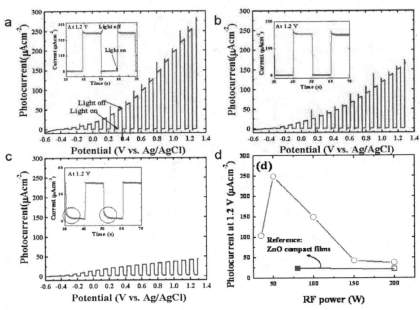

Figure 6. PEC performance of various ZnO films. **a, b, and c,** Photocurrent-voltage curves of the 50 W-, 100 W-, and 150 W-ZnO nanostructural films, respectively, measured under chopped illumination with an UV/IR cut-off filter. Insets are the current transient with time performed at 1.2 V under the light on/off illumination. **d,** Photocurrents of different ZnO films measured at 1.2 V.

The insets in Figs. 6a-6c show the current transients, performed at constant 1.2 V under light on/off illumination. The photocurrents of the ZnO nanocoral structures decay very sharply without exhibiting photocurrent tails under light-off conditions. However, the photocurrents for the nanocrystalline ZnO structures (150 W-ZnO) decays slowly showing photocurrent tails, indicating a trap-related carrier transport process (see the circles in the inset of Fig. 6c). [41, 44-46] This suggests that the ZnO nanocoral films have much better carrier transport than the nanocrystalline structures. This can be attributed to the deformation-free nature and smooth electrical pathway as shown in the SEM results. The 35 W-ZnO nanocoral structure exhibited a lower PEC response than the other nanocoral structures, indicating that the surface area of the nanoparticle films is not as large as the nanocoral films. The comparative PEC performance tests have been repeated many runs, and the results are reproducible. The greatly enhanced PEC response of the aligned nanorod and nanocoral structure has advantages for PEC applications such as water splitting by sunlight, dye-sensitized solar cells, and Li batteries.

CONCLUSIONS

In summary, we have synthesized ZnO aligned nanorod and porous ZnO nanocoral structures on FTO substrates. We found that the presence of N_2 in the growth ambient help to promote the formation of aligned nanorods at high substrate temperature of 500°C, resulting in the significantly enhanced PEC response, compared to ZnO(Ar) films deposited in pure Ar gas ambient. The nanocoral structures are composed of nanosheets, whose sizes can be controlled by the RF power for the Zn film deposition. The nanocoral structure provides a large surface area, superior light trapping, and an excellent pathway for carrier transport. Nanocoral structures have demonstrated a tenfold increase in the PEC response at 1.2 V as compared to compact ZnO films. Our results suggest that synthesis method can be used to produce desired properties of thin films and nanostructures may help to improve PEC performance.

ACKNOWLEDGEMENTS

This work was supported by the U.S. Department of Energy through the UNLV Research Foundation under Contract # DE-AC36-99-GO10337.

REFERENCES

[1] A. Fujishima and K. Honda, *Nature* (London) **238**, p.37, 1972.

[2] R. Asahi, T. Morikawa, T. Ohwaki, K. Aoki, and Y. Taga, *Science* **293**, p.269, 2001.

[3] O. Khaselev and J. A. Turner, *Science* **280**, p.425, 1998.

[4] V.M. Aroutiounian, V.M. Arakelyan, and G.E. Shahnazaryan, *Solar Energy* **78**, p.581, 2005.

[5] J. Yuan, M. Chen, J. Shi, and W. Shangguan, *Inter. J. Hydrogen Energy*, **31**, p.1326, 2006.

[6] G. K. Mor, K. Shankar, M. Paulose, O. K. Varghese, and C. A. Grimes, *Nano Lett.* **5**, p.191, 2005.

[7] B. O'Regan and M. Grätzel, *Nature*, **353**, p.737, 1991.

[8] K. Kakiuchi, E. Hosono, and S. Fujihara, *J. Photochem. & Photobiol. A: Chem.* **179**, p.81, 2006.

[9] T. F. Jaramillo, S. H. Baeck, A. Kleiman-Shwarsctein, and E. W. McFarland, *Macromol. Rapid Comm.* **25**, p.297, 2004.

[10] X. Liu, X. Wu, H. Cao, and R. P. H. Chang, *J. Appl. Phys.* **95**, p.3141, 2004.

[11] W. I. Park, D. H. Kim, S. –W. Jung, and G. –C. Yi, *Appl. Phys. Lett.* **80**, p.4232, 2002.

[12] S. Choopun, H. Tabata, and T. Kawai, *J. Cryst. Growth* **274**, p.167, 2005.

[13] F. Xu, Z. –Y. Yuan, G. –H. Du, T. –Z. Ren, C. Bouvy, M. Halasa, and B. –L. Su, *Nanotech.* **17**, p.588, 2006.

[14] K. –S. Ahn, S. Shet, T. Deutsch, C. S. Jiang, Y. Yan, M. Al-Jassim, and J. Turner, *J. Power Source*, **176**, p.387, 2008.

[15] S. Shet, K. –S. Ahn, Y. Yan, T. Deutsch, K. M. Chrusrowski, J. Turner, M. Al-Jassim, and N. Ravindra, *J. Appl. Phys.* **103**, p.073504, 2008.

[16] S. Shet, K. –S. Ahn, T. Deutsch, H. Wang, N. Ravindra, Y. Yan, J. Turner, M. Al-Jassim, *J. Mater. Research* **25**, (2010) 69 Doi: 10.1557/JMR.2010.0017.

[17] S. Shet, K.-S Ahn, T. Deutsch, H. Wang, N. Ravindra, Y. Yan, J. Turner, M. Al-Jassim, *J. Power Sources* **195**, p.5801, 2010.

[18]H. Wang, T. Deutsch, S. Shet, K. Ahn, Y. Yan, M. Al-Jassim and J. Turner, S*olar Hydrogen and Nanotechnology IV, SPIE,* Nanoscience + Engineering, p.7408, 2009.

[19]S. Shet, K. Ahn, N. Ravindra, Y. Yan, T. Deutsch, J. Turner, M. Al-Jassim, *Proceedings of the Materials Science & Technology*, p.219, 2009.

[20]S. Shet, K. Ahn, N. Ravindra, Y. Yan, T. Deutsch, J. Turner, M. Al-Jassim, *Proceedings of the Materials Science & Technology*, p.277, 2009.

[21]K.-S. Ahn, Y. Yan, S. Shet, T. Deutsch, J. Turner, and M. Al-Jassim, *Appl. Phys. Lett.* **91**, p.231909, 2007.

[22]K.-S. Ahn, Y. Yan, M.-S. Kang, J.-Y. Kim, S. Shet, H. Wang, J. Turner, and M. Al-Jassim, *Appl. Phys. Lett.* **95**, p.022116, 2009.

[23]S. Shet, K. –S. Ahn, H. Wang, N. Ravindra, Y. Yan, J. Turner, M. Al-Jassim, *J. Mater. Science* (2010) (2010) DOI 10.1007/s10853-010-4561-x.

[24]Y. Yan, K. Ahn, S. Shet, T. Deutsch, M. Huda, S. Wei, J. Turner, M. Al-Jassim, Solar Hydrogen and Nanotechnology II. Edited by Guo, Jinghua. *Proceedings of the SPIE*, **6650**, p.66500H, 2007.

[25]S. Shet, K. Ahn, N. Ravindra, Y. Yan, T. Deutsch, J. Turner, M. Al-Jassim, Materials Science & Technology 2009, *Ceramic Transactions volume*, (2010) in press.

[26]K. Ahn, S. Shet, Y. Yan, J. Turner, M. Al-Jassim, N. M. Ravindra, *Proceedings of the Materials Science & Technology*, p.901, 2008.

[27]K.-S. Ahn, Y. Yan, S. Shet, K. Jones, T. Deutsch, J. Turner, M. Al-Jassim, *Appl. Phys. Lett.* **93**, p.163117, 2008.

[28]S. Shet, K. –S. Ahn, N. Ravindra, Y. Yan, J. Turner, M. Al-Jassim, *J. Materials* **62**, p.25, 2010.

[29]K. Ahn, S. Shet, T. Deutsch, Y. Yan, J. Turner, M. Al-Jassim, N. M. Ravindra, *Proceedings of the Materials Science & Technology*, p.952, 2008.

[30]S. Shet, K. Ahn, T. Deutsch, Y. Yan, J. Turner, M. Al-Jassim, N. Ravindra, *Proceedings of the Materials Science & Technology*, p.920, 2008.

[31]K. –S. Ahn, Y. Yan, S. –H. Lee, T. Deutsch, J. Turner, C. E. Tracy, C. Perkins, and M. Al-Jassim, *J. Electrochem. Soc.* **154**, p.B956, 2007.

[32]L. Chen, S. Shet, H. Tang, H. Wang, Y. Yan, J. Turner, and M. Al-Jassim, *J. Mater. Chem.*, 2010, **20**, 6962-6967, DOI: 10.1039/c0jm01228a.

[33]L. Chen, S. Shet, H. Tang, H. Wang, Y. Yan, J. Turner, and M. Al-Jassim, *J. Appl. Phys*, **108**, 043502 (2010); doi:10.1063/1.3475714

[34]S. Shet, K.-S. Ahn, Y. Yan, N. M. Ravindra, T. Deutsch, J. Turner, M. Al-Jassim, submitted to Journal of Thin Solid films.

[35]S. –H. Kang, J. –Y. Kim, Y. Kim, H. –S. Kim, and Y. –E. Sung, *J. Phys. Chem. C* **111**, p.9614, 2007.

[36]Y. Li, X. Li, C. Yang, and Y. Li, *J. Mater. Chem.* **13**, p.2641, 2003.

[37]X. Liu, X. Wu, H. Cao, and R. P. H. Chang, *J. Appl. Phys.* **95**, p.3141, 2004.

[38]W. I. Park, D. H. Kim, S. –W. Jung, and G. –C. Yi, *Appl. Phys. Lett.* **80**, p.4232, 2002.

[39]S. Choopun, H. Tabata, and T. Kawai, *J. Cryst. Growth* **274**, p.167, 2005.

[40]F. Xu, Z. –Y. Yuan, G. –H. Du, T. –Z. Ren, C. Bouvy, M. Halasa, and B. –L. Su, *Nanotech.* **17**, p.588, 2006.

[41]M. Law, L. E. Greene, J. C. Johnson, R. Saykally, and P. Yang, *Nature Mater.* **4**,

p.455, 2005.

[42]C. M. Lopez, & K. S. Choi, *Chemical Communications*, 3328, 2005.

[43]K. Vinodgopal, S. Hotchandani, & P. V. Kamat, *Journal of Physical Chemistry* **97**, p.9040, 1993.

[44]A. Ghicov, H. Tsuchiya, J. M. Macak, & P. Schmuki, *Phys Status Solidi A* **203** p.R28, 2006.

[45]S. Nakade, T. Kanzaki, Y. Wada, & S. Yanagida, *Langmuir* **21**, p.10803, 2005.

[46]K.-S. Ahn, M. S. Kang, J. K. Lee, B. C. Shin, & J. W. Lee, *Applied Physics Letters* **89**, p.013103, 2006.

CAPPED CoFe₂O₄ NANOPARTICLES: NON-HYDROLYTIC SYNTHESIS, CHARACTERIZATION, AND POTENTIAL APPLICATIONS AS MAGNETIC EXTRACTANTS AND IN FERROFLUIDS

Tarek M. Trad[a], Rose M. Alvarez[a], Edward J. McCumiskey[b], and Curtis R. Taylor[b]

a) Department of Chemistry and Environmental Sciences, University of Texas at Brownsville and Texas Southmost College, Brownsville, Texas 78520
b) Department of Mechanical and Aerospace Engineering, University of Florida, Gainesville, Florida 32611

ABSTRACT
Magnetic cobalt ferrite nanoparticles (CoFe₂O₄) have been synthesized by a modified wet organic phase method using stable ferric and cobalt salts with octanoic and stearic acids as capping agents in anaerobic conditions. The shape, morphology, and average nanoparticle size were characterized by Transmission electron microscopy (TEM). Surface functionality of the precursor and product was determined by Fourier Transform Infra-red spectroscopy. Particle structure and crystallite size were investigated using X-ray Powder Diffraction (XRD). Surface area analysis was used to measure the specific surface area of precursors and nanoparticulate products using nitrogen adsorption. The resulting particles had an average diameter of 5 nm, were readily dispersible in organic solvents, and agglomerated in water. The ease of separation from aqueous solutions and high adsorption capacity allows for their potential utilization as magnetic extractants.

INTRODUCTION

Spinel ferrites with the general formula MFe₂O₄, where M = Mn, Co, Ni Cu, Zn, are an attractive group of magnetic materials known for their thermal and chemical stability as well as functional magnetic and electrical properties. These cubic spinels have their oxygens in an fcc close packing and their metal ions in tetrahedral or octahedral interstitial sites [1]. Ferrites are currently used in microwave devices [2], recording media [3], and magnetic resonance imaging (MRI) [4]. The majority of constituting atoms in metal ferrite nanoparticles within 1-10 nm average diameters is located at or near the surface which leads to significant changes in the material's properties rendering them promising for future highly sensitive magnetic and biomedical nanodevices. Therefore, an efficient and versatile route for synthesizing small (1-10 nm) transition metal ferrite nanoparticles with narrow size distribution is needed.

Cobalt ferrite is a particularly interesting member of the spinel ferrite family. The magnetic behavior of its nanoparticles have been extensively studied and reported [5-7]. Increased magnetic anisotropy, magnetic moment, and coercivity of CoFe₂O₄ nanoparticles compared to their magnetite counterparts play an important role in biomedical applications. It allows for the use of smaller cobalt ferrite nanoparticles for the assembly and design of smaller biocompatible devices to enhance cellular uptake and avoid the reticuloendothelial system [8]. These specific magnetic properties suggest material's enhanced efficiency for magnetic fluid hyperthermia (MFH) [8]. In addition to biomedical applications, nanoparticulate CoFe₂O₄ was investigated for its photomagnetic activity in exhibiting light-induced coercivity change [9, 10], as a heterogeneous catalyst for the oxidation of cyclohexane [11], and electrocatalytically active component towards oxygen electroreduction in composite conductive polymer/metal oxide nanoparticle electrodes [12,

[13]. Naturally, the properties are significantly influenced by the preparation method. Published techniques for the chemical synthesis of cobalt ferrite nanoparticles include hydrothermal synthesis [14, 15], sol-gel [16], sonochemical [17], micelle precipitation [18], polymerizable complex (PC) route [19], electrospinning [20], and coprecipitation [21, 22]. Aqueous coprecipitation is perhaps the most popular and versatile technique for large nanoparticle yield and reasonable product size and composition control [23].

In this article, a modified water-free co-precipitation method for the preparation of highly stable cobalt ferrite nanoparticles (~ 5 nm average particle diameter), and ferrofluid is presented. Similar route for the synthesis of γ-Fe$_2$O$_3$ nanparticles was reported [24]. CoFe$_2$O$_4$ was specifically chosen because it possesses exceptional chemical stability and mechanical strength which makes it an ideal candidate for utilization as a magnetic extractant for the remediation of arsenate and other pollutants from water.

EXPERIMENTAL

Materials

The chemicals used in this study were iron(III)nitrate nonahydrate Fe(NO$_3$)$_3$.9H$_2$O (Aldrich), cobalt(II)nitrate hexahydrate Co(NO$_3$)$_2$.6H$_2$O (Aldrich), nickel(II) nitrate hexahydrate Ni(NO$_3$)$_2$.6H$_2$O (Fluka), stearic acid C$_{18}$H$_{36}$O$_2$, caprylic acid C$_8$H$_{16}$O$_2$ (Fluka, GC grade), ammonium hydroxide solution NH$_4$OH, tetralin 1,2,3,4-Tetrahydronaphthalene C$_{10}$H$_{12}$ (Fluka), ethanol CH$_3$CH$_2$OH, and acetone (CH$_3$)$_2$CO. The chemicals were all of analytical reagent grade and were used as received and without any further purification.

Synthesis of Fe-Co(stearate) precursor

1:1 mole ratio of iron(III)nitrate nonahydrate to cobalt(II)nitrate hexahydrate was used to prepare the precursor for the cobalt ferrite nanoparticles. Iron and cobalt salts were each dissolved in 20.0 ml of ethanol, the solutions were mixed and magnetically stirred. In a separate flask, an equivalent mole ratio of stearic acid was dissolved in 70.0 ml of ethanol then added to the stirring mixture of iron and cobalt solutions. 10.0 ml of ammonium hydroxide was added dropwise to the mixture leading to the formation of a light brown precipitate. The mixture was allowed to stir for 2 hours before being filtered and washed with hot ethanol. The brownish yellow precipitate was dried in a vacuum chamber overnight. Similarly, Fe-Co(octanoate) precursor was prepared using octanoic acid instead of stearic acid.

Synthesis of the stearate-capped cobalt ferrite nanoparticles

2.0 grams of the cobalt(II), iron(III) hydroxide stearate precursor were dispersed in 100.0 ml of tetralin in a 30.0 cm long, 2.5 cm diameter reaction tube with a seal control and gas inlet. The dispersion was allowed to degas by vacuum for one hour then nitrogen gas was introduced to the system using a gas/vacuum manifold. Immediately after purging with inert gas, the reaction tube containing the yellow dispersion was placed in a tube furnace and the temperature was ramped up to 210 °C (boiling point of tetralin = 210 °C) at a ramp rate of 30 °C/min. The reaction was maintained at this temperature for 6.0 hours then left to cool to room temperature where a dark brown, ultrafine dispersion was produced. The nanoparticles were isolated from the tetraline mixture by the addition of 200.0 ml of acetone and centrifuging at 4500 rpm for 1.0 hour. The product was placed in a glass Petri dish and dried under vacuum overnight. Finally, a dark brown fine powder appeared after slowly grinding the dried product which showed immediate response to a handheld magnet. In a similar procedure, octanoate capped particles were prepared using a Fe-Co(octanoate) precursor.

Characterization methods

Synthesized precursors as well as nanoparticles were characterized using a variety of methods. Phase identification was performed using a Bruker D2Phaser powder X-ray diffractometer. 2θ values were collected from 30° to 80° using Co Kα radiation (λ = 1.788965 Å) with a current flux of 10mA and an acceleration voltage of 30 kV, a step size of 0.005 and a counting time of 5 seconds per step. All the XRD scans were collected at ambient temperature, and the phases were identified using the International Center for Diffraction Data (ICDD) database. The nature of the organic coating around nanoparticles' surface was determined by fourier transform infrared spectroscopy (FTIR) using a Nicolet Magna_IR 560 spectrometer. Standard KBr pellets were prepared using approximately 2.0 mg of sample diluted with 1.0 g of KBr. Infrared spectra in the 4000 – 400 cm^{-1} region were collected. Typically, 128 scans were recorded and averaged for each sample (4.0 cm^{-1} resolution) and the background was automatically subtracted. Transmission electron micrographs (TEM) were captured using a JEOL 2010F Field Emission TEM operated at 80 kV. For this purpose a 0.1 % w/v stable dispersions of the nanoparticles in toluene were prepared. A drop of each of the powder dispersions was carefully placed on a copper grid surface and dried before micrographs were generated. Brunauer, Emmett and Teller (BET) specific surface area measurements were performed using a NOVA 2200e surface area analyzer.

RESULTS AND DISCUSSION

Fig. 1 shows the diffraction pattern of the magnetic powder product. The positions of all XRD peaks corresponding to Bragg reflections from (220), (311), (400), (422), (511), and (440) planes match characteristic peaks of pure spinel phase cobalt ferrite, CoFe$_2$O$_4$ (ICDD-PDF# 00-022-1086). Broad XRD peaks reflections are indicative of small average nanoparticle diameter as explained by the Sherrer's equation. Broadening of peaks was also associated with the increase in non-polar solvent concentration and dielectric constant in nanoparticle synthesis [25]. The average crystallite size was obtained from the most intense peak (311) using the Debye-Scherrer formula.

$$D = \frac{K\lambda}{\beta \cos \theta}$$

Where D is the crystallite size, β the full width at half maximum (rad), λ the wavelength of the X-ray, and θ the angle between the incident and diffracted beams. The average crystallite size was found to be 55.4 Å which is very close to the average particle size determined using Transmission Electron Microscopy (TEM). Multi-point BET analysis gives a specific surface area (SSA) of 14.4 m^2/g for the Fe-Co(stearate) precursor and 1.8 m^2/g for capped cobalt ferrite nanoparticles indicating a porous precursor and an agglomerated nanoparticulate powder product. Organic coating layer and adsorbed water may be an additional cause for low product SSA.

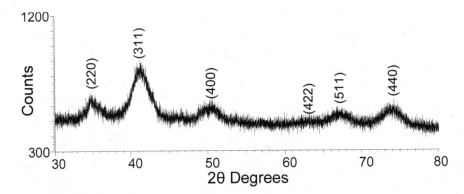

Figure 1. XRD pattern of cobalt ferrite powder sample. Spinel ferrite reflections are identified on the scan

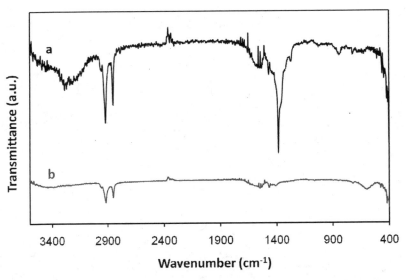

Figure 2. FTIR spectra of Co-Fe(steartate) precursor (a), and stearate capped cobalt ferrite nanoparticles (b)

FTIR spectra of the precursor (a) and nanoparticles (b) depicted in fig. 2 were carried out to confirm the presence of the capping organic moiety around the nanoparticle surface. The broad absorption band between 3220 and 3440 cm^{-1} observed in the precursor and the nanoparticle spectra can be attributed to the stretching mode of the surface hydroxyl groups. The presence of –CH$_2$– groups in the long aliphatic stearate molecules is identified by the C–H asymmetric stretching vibration at (2850–2918) cm^{-1} in fig. 2a. These bands are retained in the cobalt ferrite spectra (fig. 2b), which affirm the anchoring of the stearate molecules on nanoparticle surface. Atmospheric carbon dioxide causes the appearance of a small band around 2350 cm^{-1}. Below 2000 cm^{-1}, stretching vibration of the carboxylate group COO$^-$ is observed at ~1385 cm^{-1} while bands at ~ 1564 and 1529 cm^{-1} are attributed to asymmetric COO$^-$ stretching. Intensity of the symmetric and asymmetric C–O stretching bands decreases after precursor treatment and preparation of cobalt ferrite. Their existence, however, and the narrow separation of ~ 100 cm^{-1} between the symmetric and asymmetric carboxylate stretches suggests a chelating or both chelating and bridging modes for coordination for the stearate moieties to the surface [26]. The band at 577 cm^{-1} in fig. 2b strongly suggests the intrinsic vibrations of the metal (Fe-O) at the tetrahedral site [20], thus, the transition from an iron hydroxide nature in the precursor to iron oxide in the nanoparticle.

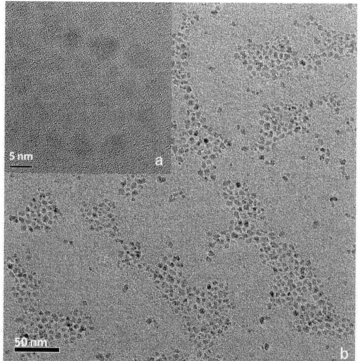

Figure 3. TEM images of CoFe$_2$O$_4$ nanoparticles at 5.0 x 10^5 magnification (a), at 5.0 x 10^4 magnification (b), and

The synthesized nanoparticles were readily dispersible in organic liquids (toluene, tetrahydrofuran, and chloroform) and stable for months. In such magnetic fluids, the organic coating confirmed by FTIR studies create steric repulsion among neighboring nanoparticles, thus, maintaining colloidal stability [27]. The outer end of the stearate molecules may be tailored to produce polar magnetic nanoparticles. On the other hand, ionic magnetic fluids depend on electrostatic repulsion to preserve colloidal stability which requires precise control of surface charge density and ionic strength otherwise nanoparticle aggregation in liquid carrier and flocculation will occur [14].

Transmission electron micrographs of the synthesized cobalt ferrite nanoparticles were used to obtain particle shape, size, and dispersity. At higher magnification, fig. 3a shows the formation of semispherical particles and fig. 3b clearly reveals uniform nanoparticle dispersion within an island formation. Narrow particle size distribution based on 150 particles and calculated average particle diameter (5.26 nm) is presented in fig 4. The Sherrer analysis of the (311) peak indicated an average diameter of 5.54 nm for the cobalt ferrite crystallites, which is in good agreement with TEM micrographs.

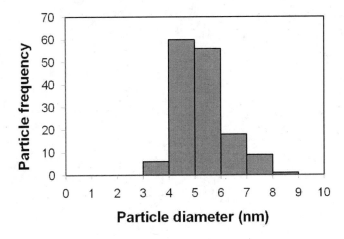

Figure 4. Particle size distribution obtained from TEM images

CONCLUSIONS

In summary, we have presented the possibility of non-hydrolytic synthesis of semi-spherical, lipophilic, and narrow particle diameter (~5 nm) cobalt ferrite nanoparticles. Due to their unique magnetic properties and small size compared to magnetite, the prepared cobalt ferrites are promising nanomaterials for biomedical applications such as MFH, and most importantly as nanoextractants. The mode of stearate chelation to the particle surface, particle hydrophobicity, and ferrite affinity to dissolved arsenate makes them ideal for utilization in arsenic remediation. The particles were shown to slowly agglomerate in solution, which would make for facile magnetic separation from solution via magnetic extraction. Careful investigations of the material's efficiency in arsenate adsorption, behavior in aqueous solutions, and particle regeneration is currently underway.

ACKNOWLEDGEMENTS

The financial support for instrument acquisition from the Serafy Foundation, Brownsville, Texas is gratefully acknowledged. The authors wish to thank Dr. Gene Paull, Chairperson of the Department of Chemistry and Environmental Sciences at UTB for his support and departmental laboratories management.

REFERENCES

[1]Sun, S. H., Zeng, H., Robinson, D. B., Raoux, S., Rice, P. M., Wang, S. X., and Li, G. X., Monodisperse MFe2O4 (M = Fe, Co, Mn) nanoparticles, *J. Am. Chem. Soc.*, **126**, (1), 273-279 (2004)

[2]Pardavi-Horvath, M., Microwave applications of soft ferrites, *Journal of Magnetism and Magnetic Materials*, **215**, 171-183 (2000)

[3]Simonds, J. L., Magnetoelectronics Today and Tomorrow, *Phys. Today*, **48**, (4), 26-32 (1995)

[4]Hogemann, D., Josephson, L., Weissleder, R., and Basilion, J. P., Improvement of MRI probes to allow efficient detection of gene expression, *Bioconjugate Chem.*, **11**, (6), 941-946 (2000)

[5]Cao, X. B., and Gu, L., Spindly cobalt ferrite nanocrystals: Preparation, characterization and magnetic properties, *Nanotechnology*, **16**, (2), 180-185 (2005)

[6]Sivakumar, N., Narayanasamy, A., Shinoda, K., Chinnasamy, C. N., Jeyadevan, B., and Greneche, J. M., Electrical and magnetic properties of chemically derived nanocrystalline cobalt ferrite, *J. Appl. Phys.*, **102**, (1), 8 (2007)

[7]Masheva, V., Grigorova, M., Valkov, N., Blythe, H. J., Midlarz, T., Blaskov, V., Geshev, J., and Mikhov, M., On the magnetic properties of nanosized CoFe2O4, *Journal of Magnetism and Magnetic Materials*, **196**, 128-130 (1999)

[8]Baldi, G., Bonacchi, D., Franchini, M. C., Gentili, D., Lorenzi, G., Ricci, A., and Ravagli, C., Synthesis and coating of cobalt ferrite nanoparticles: A first step toward the obtainment of new magnetic nanocarriers, *Langmuir*, **23**, (7), 4026-4028 (2007)

[9]Giri, A. K., Kirkpatrick, E. M., Moongkhamklang, P., Majetich, S. A., and Harris, V. G., Photomagnetism and structure in cobalt ferrite nanoparticles, *Appl. Phys. Lett.*, **80**, (13), 2341-2343 (2002)

[10]Giri, A. K., Pellerin, K., Pongsaksawad, W., Sorescu, M., and Majetich, S. A., Effect of light on the magnetic properties of cobalt ferrite nanoparticles, *IEEE Trans. Magn.*, **36**, (5), 3029-3031 (2000)

[11]Tong, J. H., Bo, L. L., Li, Z., Lei, Z. Q., and Xia, C. G., Magnetic CoFe2O4 nanocrystal: A novel and efficient heterogeneous catalyst for aerobic oxidation of cyclohexane, *J. Mol. Catal. A-Chem.*, **307**, (1-2), 58-63 (2009)

[12]Singh, R. N., Lal, B., and Malviya, M., Electrocatalytic activity of electrodeposited composite films of polypyrrole and CoFe2O4 nanoparticles towards oxygen reduction reaction, *Electrochim. Acta*, **49**, (26), 4605-4612 (2004)

[13]Singh, R. N., Singh, N. K., and Singh, J. P., Electrocatalytic properties of new active ternary ferrite film anodes for O-2 evolution in alkaline medium, *Electrochim. Acta*, **47**, (24), 3873-3879 (2002)

[14]Morais, P. C., Garg, V. K., Oliveira, A. C., Silva, L. P., Azevedo, R. B., Silva, A. M. L., and Lima, E. C. D., Synthesis and characterization of size-controlled cobalt-ferrite-based ionic ferrofluids, *Journal of Magnetism and Magnetic Materials*, **225**, (1-2), 37-40 (2001)

[15]Cabanas, A., and Poliakoff, M., The continuous hydrothermal synthesis of nano-particulate ferrites in near critical and supercritical water, *J. Mater. Chem.*, **11**, (5), 1408-1416 (2001)

[16]Lee, J. G., Park, J. Y., and Kim, C. S., Growth of ultra-fine cobalt ferrite particles by a sol-gel method and their magnetic properties, *J. Mater. Sci.*, **33**, (15), 3965-3968 (1998)

[17]Shafi, K., Gedanken, A., Prozorov, R., and Balogh, J., Sonochemical preparation and size-dependent properties of nanostructured CoFe2O4 particles, *Chem. Mat.*, **10**, (11), 3445-3450 (1998)

[18]Jeppson, P., Sailer, R., Jarabek, E., Sandstrom, J., Anderson, B., Bremer, M., Grier, D. G., Schulz, D. L., Caruso, A. N., Payne, S. A., Eames, P., Tondra, M., He, H. S., and Chrisey, D. B., Cobalt ferrite nanoparticles: Achieving the superparamagnetic limit by chemical reduction, *J. Appl. Phys.*, **100**, (11), 7 (2006)

[19]Emamian, H. R., Honarbakhsh-raouf, A., Ataie, A., and Yourdkhani, A., Synthesis and magnetic characterization of MCM-41/CoFe2O4 nano-composite, *J. Alloy. Compd.*, **480**, (2), 681-683 (2009)

[20]Sangmanee, M., and Maensiri, S., Nanostructures and magnetic properties of cobalt ferrite (CoFe2O4) fabricated by electrospinning, *Applied Physics a-Materials Science & Processing*, **97**, (1), 167-177 (2009)

[21]Grigorova, M., Blythe, H. J., Blaskov, V., Rusanov, V., Petkov, V., Masheva, V., Nihtianova, D., Martinez, L. M., Munoz, J. S., and Mikhov, M., Magnetic properties and Mossbauer spectra of nanosized CoFe2O4 powders, *Journal of Magnetism and Magnetic Materials*, **183**, (1-2), 163-172 (1998)

[22]Chinnasamy, C. N., Senoue, M., Jeyadevan, B., Perales-Perez, O., Shinoda, K., and Tohji, K., Synthesis of size-controlled cobalt ferrite particles with high coercivity and squareness ratio, *J. Colloid Interface Sci.*, **263**, (1), 80-83 (2003)

[23]Olsson, R. T., Salazar-Alvarez, G., Hedenqvist, M. S., Gedde, U. W., Lindberg, F., and Savage, S. J., Controlled synthesis of near-stoichiometric cobalt ferrite nanoparticles, *Chem. Mat.*, **17**, (20), 5109-5118 (2005)

[24]Bourlinos, A. B., Simopoulos, A., and Petridis, D., Synthesis of Capped Ultrafine g-Fe2O3 Particles from Iron(III) Hydroxide Caprylate: A Novel Starting Material for Readily Attainable Organosols, *Chem. Mat.*, **14**, (2), 899-903 (2002)

[25]Ayyappan, S., Philip, J., and Raj, B., A facile method to control the size and magnetic properties of CoFe2O4 nanoparticles, *Mater. Chem. Phys.*, **115**, (2-3), 712-717 (2009)

[26]Deacon, G. B., and Phillips, R. J., Relationships between the carbon-oxygen stretching frequencies of carboxylato complexes and the type of carboxylate coordination, *Coordination Chemistry Reviews*, **33**, (3), 227-50 (1980)

[27]Schwuger, M., Haegel, F., and Buske, N., Application of magnetite sols in environmental technology. In *Surfactants and Colloids in the Environment*, Springer Berlin / Heidelberg: 1994; Vol. 95, pp 175-180.

NANOMATERIAL FIBER OPTIC SENSORS IN HEALTHCARE AND INDUSTRY APPLICATIONS

K. Sun, N. Wu, C. Guthy, and X. Wang
Department of Electrical and Computer Engineering
University of Massachusetts, Lowell, MA 01854 USA

ABSTRACT

Nanomaterials, being governed by quantum mechanical effects because of their small size, have many unique physical properties that are not present in their bulk counterparts. For optical sensing applications, the most important characteristics of nanomaterials are their discrete energy levels in semiconductor nanostructures and surface plasmon resonance in metallic nanoparticles. These characteristics dramatically change materials' optical properties, especially their absorption and emission spectra. When used in conjunction with optical fibers, nanomaterials such as metal nanoparticles and carbon nano-tubes, can dramatically improve the performance and functionality of optical fiber sensors. This paper reviews applications of these nanomaterial fiber optic sensors in structural health and environmental monitoring as well as biomedical imaging.

INTRODUCTION

Over the years, optical sensors have been developed and utilized for a wide variety of sensing applications. Compared to electrical sensors, optical sensors have many unique advantages, including immunity to electromagnetic interference[1] and ability to be multiplexed [2]. Optical fiber sensors have even more advantages over electrical sensors, such as smaller size, low cost, and remote sensing, control, and fast data transfer capabilities. With the advent of nanotechnology, nanomaterials have attracted more and more interest in their potential utility in sensing applications[3]. At the nanometer level, most materials exhibit physical and chemical properties that are radically different from their bulk counterparts. When combined with optical fiber technology, the unique properties of nanomaterials make novel nanomaterial optical fiber sensors possible. These sensors have great potential in medical and industrial applications, including environmental monitoring and biomedical imaging. This paper reviews three kinds of nanomaterial optical fiber sensors: Fabry-Perot interference-based sensors, Surface-Enhanced Raman Scattering (SERS)-based sensors, and photoacoustic sensors. For each kind of sensor, the sensing mechanism, current developments, and related research activities at the University of Massachusetts Lowell are presented.

FP interference-based sensors

The Fabry-Perot (FP) interferometer is comprised of two parallel mirrors separated by a distance L. When a beam of light travels through the cavity, it hits one of the two mirrors, and starts reflecting back and forth between the two mirrors. The resulting interference fringes are determined by the cavity length L and the refractive index n of the material inside the cavity. If one of the mirrors is fixed, and the other mirror can be displaced or deflected by an input stimulus, or if the refractive index changes with external environmental variables or physical/chemical bonding processes, the optical path of FP cavity changes and interference fringes will shift accordingly. FP optical sensors have been used to sense temperature[4, 5], strain[6], pressure[7-9], sound[10], and many other physical variables.

Figure 1 is a diagram of a FP interferometer-based optical sensor developed at the University of Massachusetts Lowell made with surface-arranged single-walled carbon nanotubes (SWCNTs). SWCNTs are bundles of nanotubes, each between 20 to 40 nm in diameters and several microns long. They are an ideal material for sensing applications that require sensitivity because of their hollow structure, high surface area, and high chemical reactivity [11]. The endface of a single-mode optical fiber is coated with a layer of these SWCNTs, forming an FP cavity between the endface of the fiber and the top surface of SWCNT layer. Whenever the adsorption of the targeted analyte into the SWCNT layer changes the refractive index of SWCNT layer, the reflectance spectra shift.

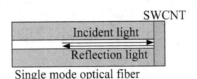

Figure 1. A FP cavity-based sensor with a thin layer of SWCNT on the distal end of a single-mode fiber.

To make the solution needed to coat SWCNTs onto the endface of an optical fiber, ten milligrams of purified SWCNTs (synthesized by High-Pressure CO conversion) are mixed with 100 milligrams of Triton X-100 ($C_{14}H_{22}O(C_2H_4O)_n$, a commercial nonionic surfactant) in 15 milliliters of deionized water, forming a SWCNT dispersion. An uncoated optical fiber is then dipped into the SWCNT dispersion for 20 seconds. After being pulled out of the dispersion, the SWCNT-coated fiber tip is dried with a heat gun. The whole procedure can be repeated as many times as necessary, depending on the number of coatings required for the sensing application.

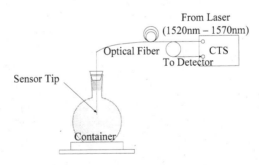

Figure 2. Sensing DMMP vapor with a SWCNT Fabry-Perot optical fiber sensor.

Figure 2 depicts a typical sensing setup using a SWCNT FP optical fiber sensor[12]. The sensor is placed inside a 100 milliliter round-bottom flask containing 30 milliliters of dimethyl methylphosphonate (DMMP, a simulant for the nerve agent Soman) and coupled to a component testing system (CTS) (Micro Optics Inc. Si720). The CTS's internal tunable laser has a wavelength range between 1520 nm and 1570 nm. The dynamic sensing test results are shown in Figure 3. The reflectance peak shifting with the dose of DMMP is very linear, and the sensor response time is fast (less than 30 seconds). The sensitivity is calculated as 7.2×10^{-3} dB/ppm.

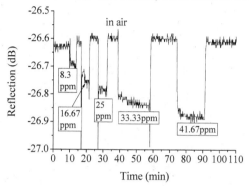

Figure 3. Reflectance spectrum peak shifts at different dose of DMMP.

SERS-based sensors

Surface-enhanced Raman scattering (SERS) has attracted much attention since its discovery in the 1974[13]. It has been observed that noble metal nanoparticles can greatly enhance the Raman scattering signal of the molecules adsorbed on the nanoparticle surface. This enhancement is believed to be caused by the excitation of localized surface plasmon[14]. For Rhodamine 6G molecules on nanoparticle SERS substrate, the enhancement factors (EFs) can be up to $10^{14} \sim 10^{15}$, making it possible to make an extremely sensitive sensor that can detect individual single molecules[15]. A variety of fabrication techniques have been developed for high-EF SERS optical fiber probes [16-18], many of which techniques involve roughening the tip of an optical fiber, then coating it with a thin layer of noble metal nanoparticles, such as silver or gold.

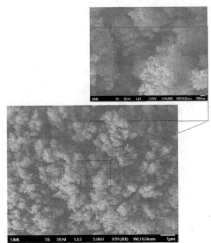

Figure 4. SEM photo of the nanospikes on the silica ablated by femtosecond laser.

Researchers at the University of Massachusetts Lowell successfully fabricated high-EF SERS probes on optical fiber tips using a femtosecond (fs) laser micromachining technique[19]. This technique, an fs laser scanning process, created a micrograting-like structure with textured nano-spikes (Figure 4).The laser-ablated fiber endface was then SERS activated by thermally depositing silver on it. As seen in Figure 5, when excited by an excitation laser, the sensor detected a high-quality SERS signal from a solution of Rhodamine 6G.

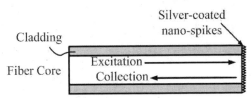

Figure 5. A layer of silver-coated silica nanospikes acts as an efficient SERS substrate.

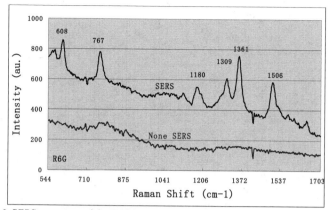

Figure 6. SERS spectrum of the R6G is greatly enhanced by the roughened, silver-coated silica substrate with nanospikes.

Figure 6 compares Raman spectrum signals obtained on two surfaces: one with nanospikes, and a reference surface without nanospikes. The strength of the Raman scattering was found to be greatly enhanced because of the silver-coated nanospikes.

Photoacoustic sensors

Photoacoustic imaging (PAI) is a non-invasive imaging method that combines the spectral selectivity of molecular excitation by laser irradiation with the high resolution of ultrasound imaging[20]. The operating principle behind PAI is the photothermal effect. As seen in Figure 7, when the excitation laser irradiates a sample tissue, a certain amount of the laser's energy is absorbed by the tissue and converted into heat. The resulting thermal expansion causes the tissue to generate ultrasonic signals. These signals can be collected by an ultrasound transducer and used to analyze the tissue. While certainly an interesting and useful diagnostic tool, the PAI system is not without its faults. If the minimum absorption wavelength of a tissue is within the PAI's operating wavelength range, the tissue becomes opaque to the PAI system. At the other extreme, if most of the laser energy is absorbed on the surface of the tissue, the PAI system can only probe the tissue's surface.

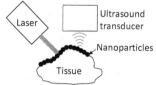

Figure 7. Operating Principle of Photoacoustic Imaging

Nanoparticles (nanorods, nanospheres, carbon nanotube, nanoshells, etc.) can resolve the above mentioned problems that affect the functionality of a PAI system. PAI probes coated with biocompatible nanoparticles can be treated with biolinker-like antibodies that immobilize on certain tissues, improving the probes' selectivity. The nanoparticles also help the PAI system probe deeper into the tissue and even probe the previously opaque tissues. Since the characteristic absorption peaks of nanoparticles are dependent on the size of the nanoparticle being used, it is possible to choose suitable nanoparticles for the application so that its absorption bands match the wavelength of the PAI system's laser. This enhances photoacoustic generation of the probes, ensuring that good ultrasonic images can be obtained.

Lihong Wang's group used gold nanoshells as in-vivo contrast agents to image rat brains [21-27]. Each nanoshell were made of a silica core with a diameter of 125 nm and coated with a 10-12 nm thick gold layer, and had a peak absorption wavelength of 800 nm. Since most tissue absorbs very little energy at this wavelength, they appeared opaque to Wang's PAI probe. When the probe was excited by an 800 nm laser, it was able to distinguish normal tissue from tissue infused with nanoshells because of the difference of absorption efficiency at this wavelength. The nanoshells had a 3.7 hour half life time in the bloodstreams of immune-competent BALB/C mice. Photoacoustic tomography images showed clear pictures of the rat brain's vessels.

Vladimir P. Zharov's group used gold-coated carbon nanotubes (GNTs) as a contrast agent for PAI [28]. In their research, the GNTs were comprised of a shortened single-wall carbon nanotube core with a diameter of 1.5-2 nm, and were coated with a 4-8 nm thick layer of gold. The GNTs' absorption peak occurred between 520-530 nm in wavelength (similar to gold nanospheres), and had a longitudinal resonance peak in the near-infrared region near 850 nm (similar to gold nanorods). Photoacoustic signals from clustered GNTs with an average size of 250-300 nm were 10-15 times stronger than those from similarly sized carbon nanotube clusters.

The photoacoustic mechanism also allows the ultrasound generator to achieve a very high central frequency by shortening the optical irradiation pulse width. The ultrasound generation efficiency is determined by the energy absorption layer used in the photoacoustic generation process. An ideal energy absorption material would have a high absorption coefficient at the optical radiation wavelength, a high thermal expansion coefficient, and high temperature survivability. Biagi's group [29, 30] showed that, by mixing graphite in epoxy and coating it onto an optical fiber, an ultrasound signal with a 40 MHz central frequency could be generated.

CONCLUSIONS

This paper reviewed three kinds of nanomaterial optical fiber sensors: Fabry-Perot interference-based sensors, surface-enhanced Raman scattering (SERS)-based sensors, and photoacoustic sensors. For each sensor, the sensing mechanism, current developments, and related research activities at University of Massachusetts Lowell were presented.

REFERENCES

[1] K.T.V. Grattan, T. Sun, Fiber optic sensor technology: an overview, Sensors and Actuators A: Physical, 82 (2000) 40-61.
[2] A.D. Kersey, T.A. Berkoff, W.W. Morey, Multiplexed fiber Bragg grating strain-sensor system with a fiber Fabry-Perot wavelength filter, Opt. Lett., 18 (1993) 1370-1372.
[3] J. Shi, Y. Zhu, X. Zhang, W.R.G. Baeyens, A.M. García-Campaña, Recent developments in nanomaterial optical sensors, TrAC Trends in Analytical Chemistry, 23 (2004) 351-360.

[4] C.E. Lee, H.F. Taylor, Fiber-optic Fabry-Perot temperature sensor using a low-coherence light source, Lightwave Technology, Journal of, 9 (1991) 129-134.

[5] R.A. Wolthuis, G.L. Mitchell, E. Saaski, J.C. Hartl, M.A. Afromowitz, Development of medical pressure and temperature sensors employing optical spectrum modulation, Biomedical Engineering, IEEE Transactions on, 38 (1991) 974-981.

[6] K.A. Murphy, M.F. Gunther, A.M. Vengsarkar, R.O. Claus, Quadrature phase-shifted, extrinsic Fabry-Perot optical fiber sensors, Opt. Lett., 16 (1991) 273-275.

[7] B. Hälg, A silicon pressure sensor with a low-cost contactless interferometric optical readout, Sensors and Actuators A: Physical, 30 (1992) 225-230.

[8] W. Wang, N. Wu, Y. Tian, X. Wang, C. Niezrecki, J. Chen, Optical pressure/acoustic sensor with precise Fabry-Perot cavity length control using angle polished fiber, Opt. Express, 17 (2009) 16613-16618.

[9] W. Wang, N. Wu, Y. Tian, C. Niezrecki, X. Wang, Miniature all-silica optical fiber pressure sensor with an ultrathin uniform diaphragm, Opt. Express, 18 (2010) 9006-9014.

[10] K. Onur, et al., External fibre Fabry–Perot acoustic sensor based on a photonic-crystal mirror, Measurement Science and Technology, 18 (2007) 3049.

[11] M. Consales, S. Campopiano, A. Cutolo, M. Penza, P. Aversa, G. Cassano, M. Giordano, A. Cusano, Carbon nanotubes thin films fiber optic and acoustic VOCs sensors: Performances analysis, Sensors and Actuators B: Chemical, 118 (2006) 232-242.

[12] N. Wu, W. Wang, X. Wang, DMMP (Dimethylmethylphosphonate) Gas Detection with Carbon Nanotube Coated Optical Fiber Sensor, in: The Fifth International Workshop on Advanced Smart Materials and Smart Structures Technology, Northeastern University, Boston, MA, USA, 2009.

[13] M. Fleischmann, P.J. Hendra, A.J. McQuillan, Raman spectra of pyridine adsorbed at a silver electrode, Chemical Physics Letters, 26 (1974) 163-166.

[14] E. Smith, G. Dent, Modern Raman Spectroscopy: A Practical Approach, John Wiley and Sons, 2005.

[15] S. Nie, S.R. Emory, Probing Single Molecules and Single Nanoparticles by Surface-Enhanced Raman Scattering, Science, 275 (1997) 1102-1106.

[16] D.L. Stokes, T. Vo-Dinh, Development of an integrated single-fiber SERS sensor, Sensors and Actuators B: Chemical, 69 (2000) 28-36.

[17] Y. Han, X. Lan, T. Wei, H.-L. Tsai, H. Xiao, Surface enhanced Raman scattering silica substrate fast fabrication by femtosecond laser pulses, Applied Physics A: Materials Science & Processing, 97 (2009) 721-724.

[18] X. Lan, Y. Han, T. Wei, Y. Zhang, L. Jiang, H.-L. Tsai, H. Xiao, Surface-enhanced Raman-scattering fiber probe fabricated by femtosecond laser, Opt. Lett., 34 (2009) 2285-2287.

[19] W. Wang, H. Huo, N. Wu, M. Shen, C. Guthy, X. Wang, Surface-enhanced Raman scattering on optical material fabricated by femtosecond laser, Proceeding of SPIE, 7673 (2010).

[20] X. Yang, E.W. Stein, S. Ashkenazi, L.V. Wang, Nanoparticles for photoacoustic imaging, Wiley Interdisciplinary Reviews: Nanomedicine and Nanobiotechnology, 1 (2009) 360-368.

[21] Y. Wang, X. Xie, X. Wang, G. Ku, K.L. Gill, D.P. O'Neal, G. Stoica, L.V. Wang, Photoacoustic Tomography of a Nanoshell Contrast Agent in the in Vivo Rat Brain, Nano Letters, 4 (2004) 1689-1692.

[22] G. Ku, L.V. Wang, Deeply penetrating photoacoustic tomography in biological tissues enhancedwith an optical contrast agent, Opt. Lett., 30 (2005) 507-509.

[23] G. Ku, X. Wang, X. Xie, G. Stoica, L. Wang, Imaging of tumor angiogenesis in rat brains in vivo by photoacoustic tomography, Appl. Opt., 44 (2005) 770-775.

[24] V. Ntziachristos, J. Ripoll, L.V. Wang, R. Weissleder, Looking and listening to light: the evolution of whole-body photonic imaging, Nature Biotechnology, 23 (2005) 313(318).

[25] H.F. Zhang, K. Maslov, G. Stoica, L.V. Wang, Functional photoacoustic microscopy for high-resolution and noninvasive in vivo imaging, Nat Biotech, 24 (2006) 848-851.

[26] X. Yang, S.E. Skrabalak, Z.-Y. Li, Y. Xia, L.V. Wang, Photoacoustic Tomography of a Rat Cerebral Cortex in vivo with Au Nanocages as an Optical Contrast Agent, Nano Letters, 7 (2007) 3798-3802.

[27] K.H. Song, C. Kim, C.M. Cobley, Y. Xia, L.V. Wang, Near-Infrared Gold Nanocages as a New Class of Tracers for Photoacoustic Sentinel Lymph Node Mapping on a Rat Model, Nano Letters, 9 (2008) 183-188.

[28] J.-W. Kim, E.I. Galanzha, E.V. Shashkov, H.-M. Moon, V.P. Zharov, Golden carbon nanotubes as multimodal photoacoustic and photothermal high-contrast molecular agents, Nat Nano, 4 (2009) 688-694.

[29] E. Biagi, F. Margheri, D. Menichelli, Efficient laser-ultrasound generation by using heavily absorbing films as targets, Ultrasonics, Ferroelectrics and Frequency Control, IEEE Transactions on, 48 (2001) 1669-1680.

[30] E. Biagi, M. Brenci, S. Fontani, L. Masotti, M. Pieraccini, Photoacoustic Generation: Optical Fiber Ultrasonic Sources for Non-Destructive Evaluation and Clinical Diagnosis, Optical Review, 4 (1997) 481-483.

PLASMONIC SILVER NANOPARTICLES FOR ENERGY AND OPTOELECTRONIC APPLICATIONS

Haoyan Wei
Institute for Shock Physics, Washington State University
Pullman, WA 99164, USA

ABSTRACT

Metal nanoparticle loaded composite thin films exhibit unique optical extinction originating from collective charge oscillations (surface plasmon) at the metal-dielectric interface induced by the incident light. Utilizing vapor phase co-deposition, we present significant progress in tailoring the optical, electrical, and electronic properties of silver nanoparticles embedded in dielectric and semiconducting polymers as a function of nanoparticle morphology by varying the metal composition and film thickness. Near the percolation threshold, the presence of silver nanoclusters with arbitrary sizes and shapes enable the unusual broadband absorption from visible to infrared regimes. This is of particular interest for developing multi-spectral photodetectors. At a relatively lower concentration, the absorption is tailored to closely match the solar spectrum, which could potentially overcome the limited wavelength response of present photovoltaic materials adversely affecting the cell efficiency. The electrical conductivity of the resulted nanocomposites can be depicted within the scope of three-stage percolation model with a significantly stretched transition. The presence of vast interfaces in nanoparticle loaded blend films leads to much stronger chemical interactions between metals and semiconducting polymers. This results in distinct interfacial electronic structures in contrast to layered films with extended planar interfaces. By varying metal composition, it is able to tune the energy levels in the blend film, which would have dramatic impact on the charge-transfer process upon photoexcitation.

INTRODUCTION

Considerable interests in nanostructured metallic particles originate from their remarkable light absorption and extinction absent in their bulk constituents. The interaction of incident electromagnetic radiation with free electrons in metal nanoparticles generates collective charge oscillations, resulting in surface plasmon polaritons at the metal-dielectric interface. The surface plasmon resonance is strongly dependent on metal species, particle geometry as well as dielectric environment.[1-4] Their versatile and tunable optical characteristics are particularly suitable for many photonic applications such as surface enhanced Raman scattering (SERS),[5,6] color filters,[7-9] and all optical switching.[10] The resonance frequency shifts to longer wavelength as the particles size enlarges. As coalescence takes place among initially isolated metallic nanoparticles, nanoclusters with more irregular shapes and broader size distributions form,[11,12] resulting in fractal structures which can greatly extend the absorption from the visible spectral range into the infrared (IR).[13-15]

One advantage of surface plasmons is the intense electromagnetic field density induced around metallic nanoparticles, which effectively concentrates light into small regions. This has been very important for thin devices such as photovoltaic cells[16-18] and light-emitting diodes (LEDs).[19,20] Their efficiency can be greatly enhanced due to strong coupling of the light to the active interface (absorption or emission region, respectively) with substantially less materials which in turn dramatically reduce the total device cost. In addition, the giant electromagnetic

171

field from surface plasmon can result in photoelectric emission from metallic nanoparticles. This mechanism has stirred interests in recent studies in the direct employment of metallic nanoparticles as possible photosensitizers.[16,21]

Optical enhancement phenomena have been studied by many research groups. However, the strong coupling of optical and electrical oscillation has been less leveraged, which has the potential merging photonics and electronics at nanoscale. In recent research efforts, we demonstrate controlled tailoring of the optical behavior of silver nanoparticles embedded in Teflon AF matrix by varying the metal concentration, size and shape. As the metal concentration is near the percolation threshold, silver nanoclusters are present in various arbitrary sizes and shapes. Unusual broadband absorption spectra from visible to infrared regimes (400 nm – 2500 nm and beyond) is enabled in these nanocomposites, making them potential candidates for developing multi-color photodetectors. At a relatively lower silver loading, we also show that the absorption profile of Ag/Teflon AF nanocomposites can be altered to resemble the solar radiation spectrum, which is very promising in the development of highly efficient photovoltaic cells.

In this work, our recent research efforts on silver/polymer nanocomposites are summarized together to help better understand the interconnection between different aspects of the materials for potential optoelectronic applications. We begin with a basic theoretical introduction of surface plasmons. Then the discussion is focused on the nanocomposite fabrication through vapor phase co-deposition of silver nanostructures on/in dielectric substrates. The characteristic microstructures are compared between pure silver particle films and their counterparts embedded in the polymer matrix. Finally, the optical, electrical, and electronic behaviors of the nanocomposites are investigated and discussed for potential photoelectric utilities. These properties are closely related to the morphology of the embedded silver nanostructures which are dependent upon both film thickness and metal composition.

SURFACE PLASMON EXCITATION OF METALLIC NANOPARTICLES

In this section, we will explain the basic physical origin of surface plasmon effect as simple as possible with the two widely accepted theories: the Drude-Maxwell model and Mie theory, in order to elucidate its application in energy and optical sensors discussed later, and leave out the complicated mathematical expression. Readers who are interested in this aspect are directed to the references listed at the end for further information.[1,3,22-24]

Plasmonic excitation is a physical behavior between incident electromagnetic radiation and free electrons in a good conductor such as metals or doped semiconductors. Metals are typically treated using Drude model where they are considered as an immobile and periodic cationic lattice filled with free mobile electrons. Under the alternating electric field of the external electromagnetic waves, the electron clouds are enabled to oscillate with respect to the cationic grid, leading to periodic oscillations of charge density at a resonant frequency, termed plasma frequency ω_p. This characteristic frequency is given by the following relation for a bulk metal:

$$\omega_p^2 = ne^2 / \varepsilon_0 m_e \tag{1}$$

Where n is the number density of electrons, e is the electron charge, ε_0 is the dielectric constant of vacuum, and m_e is the electron effective mass, respectively. Most metals have ω_p in the ultraviolet regime while some metals, such as copper, exhibit a visible plasma frequency. The quasi-particles, resulting from quantized plasma oscillations, are called plasmons.

As the dimension of a metal diminishes below 100 nm, its electronic behavior changes dramatically as a result of the increased surface-to-volume ratio. The system boundaries become very important in contrast to their bulk counterpart. For smaller particles (< 2 nm) or clusters, the quantum effects are dominant and are not the focus in the discussion. For metallic nanoparticles larger than 2 nm, free electrons are geometrically confined in a finite small volume of nanoscale, which is much smaller than the wavelength of incident light. Because of the miniature size, electrons in the same nanoparticle see almost identical electric field from the incident radiation. This induces the collective oscillation of negative charges coherently in response to the electric field of the incident light, so called localized surface plasmon excitation. The mobile electrons are displaced out of the cationic network, resulting in excess negative charges on one side of the particle surface. Consequently, excess positive charges emerge on the other surface side due to the absence of electrons (Figure 1a). The displaced electrons tend to be pulled back under the restoring force due to charge repartition. Because of the small size of the nanoparticle, the emergence of different polarity on opposite sites forms an effective electric dipole imposed on the displaced electrons, which is different from the free plasmons occurring in the bulk metals. The induced electric-dipole excitation has strong coupling with the incident electromagnetic waves, leading to surface plasmon polaritons at the metal/dielectric interface.

Mie theory, based on the scattering approach, has generally been successful in explaining the optical absorption of metallic nanoparticles. If only the dipole oscillation is assumed to contribute to the absorption, the total extinction coefficient has the following expression:

$$C_{ext} = \frac{24\pi^2 R^3 \varepsilon_m^{3/2}}{\lambda} \frac{\varepsilon''}{(\varepsilon' + 2\varepsilon_m)^2 + \varepsilon''^2} \qquad (2)$$

Where C_{ext} is the extinction cross section of a single particle, R is the particle radius, ε_m is the medium dielectric function, λ is the wavelength of the light, ε' and ε'' represent the real and imaginary parts of the material dielectric function. The resonance condition for surface plasmon excitation occurs when the C_{ext} reaches its maxima if the condition of $\varepsilon' = -2\varepsilon_m$ is satisfied. From Equation 2, it is clearly indicated that the resonance frequency is correlated with dielectric medium properties and particle size. In addition, the size effect is also manifested in the materials dielectric function which is size-dependent. The dependence of resonance frequency on particle shape can be modeled through an extension of Mie's theory by incorporating depolarization factors of different axis.

The free electrons in some noble metals such as gold, silver and copper have the appropriate density value to give the surface plasmon peaks in the visible light. For example, strong absorptions are observed typically at about 510 nm, 400 nm and 570 nm for gold, silver and copper respectively. This, along with their synthesis availability and stability, has explained the extensive studies on these materials and their applications.

The strong dependence of absorption maxima on particle size indicates the color variation of nanoparticles. Figure 1b illustrates silver nanoparticles with different size distribution on glass substrate deposited via thermal evaporation. Optical appearance of distinct color indicates clearly the correlation of surface plasmon resonance frequency with the particle size. One fascinating example of this size-sensitive coloration is the aesthetic utility such as the coloration of glass and pottery.[25]

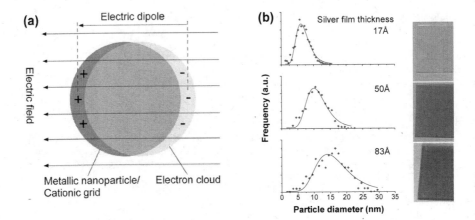

Figure 1. (a) Schematic representation of surface plasmon excitation in metallic nanoparticles as a result of displacement of free electrons from the lattice of positive nuclei by the incident light. (b) The distinct optical appearance of silver nanoparticles (right column) as a result of different plasmon resonance frequency originating from various sizes (left column).

VAPOR PHASE CO-DEPOSITION OF SILVER NANOSTRUCTURES

Wet-chemistry has been widely adopted to synthesize silver nanoparticles with a variety of sizes and shapes. The particles are then assembled after synthesis onto/into specific supporting materials according to different application requirements. As an alternative approach, vapor phase co-deposition in a high-vacuum consolidates the multi-process chemical synthesis and assembly into a single step to make particle films. We have studied the deposition of a variety of dielectric and semiconducting polymeric materials for potential optoelectronic applications. Polymers are preferably selected as matrix materials due to their low cost and excellent processability. Two prominent polymers, Teflon AF 2400 and P3HT, will be discussed in this work. Teflon AF 2400 (poly[4,5-difluoro-2,2-bis(trifluoromethyl)-1,3-dioxole-co-tetrafluoroethylene]) is an amorphous fluoropolymer (AF) which possesses excellent optical clarity and high transparency (>95%) in the visible and near-infrared wavelength ranges combined with good mechanical and chemical stability due to its amorphous nature. The chemical structure of the Teflon AF 2400 is illustrated in Figure 2a, which is composed of 13 mol% tetrafluoroethylene (TFE) monomers and 87 mol% dioxole monomers. Poly (3-hexylthiophene) (P3HT) is selected due to its high drift mobility (up to 0.1 cm^2 V^{-1} s^{-1}) and wide applications as a hole-transporting agent in developing plastic electronics and organic solar cells. Its structure is also provided in Figure 2a. The evaporizability of Teflon AF and P3HT has been demonstrated in previous studies with Fourier transform infrared spectroscopy (FTIR).[26-29]

Ag and Ag/polymer nanocomposite films were fabricated via vapor phase co-deposition in a high-vacuum chamber with a base pressure of 10^{-7} torr. A schematic drawing of the experimental setup is illustrated in Figure 2b. The synthesis chamber is equipped with an electron-beam evaporator (Mantis Deposition Ltd.) which has four individually controlled pockets, allowing for sequential or simultaneous thermal evaporation of up to four different materials. The deposition substrate holder is on the top of the chamber with an integrated resistive heater to enhance the adsorption of silver at an elevated temperature of 80~120°C. A quartz crystal microbalance (QCM) from Inficon was used to gauge the film deposition rate and thickness. The deposition of polymers are kept at a constant low deposition rate within 1~6 Å/min and the silver deposition rate was varied to alter the metal loadings.

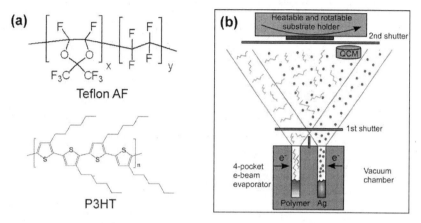

Figure 2. (a) The chemical structure of Telfon AF fluoropolymer and P3HT. (b) Schematic representation of the experimental setup for Ag and Ag loaded nanocomposite films via vapor phase co-deposition.

MICROSTRUCTURES OF SILVER NANOPARTICLE (LOADED) FILMS

Pure silver nanoparticle films are appropriate references to help understand their behavior when embedded in polymer matrix. The growth of silver nanoparticle films typically follows a tri-stage journey with film thickness: isolated island particles, interconnected clusters, and a continuous film.[30] Corresponding representative microstructures of silver films on Si substrates are shown in Figures 3a-c, which were imaged using a field emission scanning electron microscope (FESEM). Since metals do not have good wetting with dielectrics, their atoms tend to coalesce into individual isolated islands at the initial deposition to minimize the surface free energy. Figure 3a shows the island growth (Volmer–Weber mechanism)[31] in a 50Å thick film. The particle size is not uniform and exhibits a log-normal distribution as indicated in Figure 1b. At film thickness of 17Å, 50Å and 83Å, the particles presumed in round shapes have an average size in diameter of about 6–7 nm, 10-12 nm and 17 nm (Figure 1b), and an area filling fraction of about 35%, 45% and 62%, respectively. As the deposition continues, the islands grow in all three dimensions, leading to the increase of both the particle size and the size distribution. At the same time, inter-particle gaps decrease continuously as a result of the lateral enlargement of islands. The broadening of the size distribution implies that coalescence occurs in neighboring particles as they approach to each other. This leads to a growth towards fractal structures. In Figure 3b, the film has a thickness of 161Å and an area filling fraction of ca. 73%. Multiple, previously isolated islands evolve into larger irregular-shaped nanoclusters, indicating a lot of on-going interconnection activities. At this point, the film grows into the 2nd percolation stage where a quasi-continuous film forms. As the deposition continues, the substrate is fully covered with silver films which show morphology analogous to its bulk counterpart (Figure 3c).

Figures 3d-f illustrate electron micrographs of Ag/Teflon AF nanocomposites with 73 vol% Ag nanoparticle loadings at representative film thickness. Compared with individual Ag evaporation, the formation of metal nanoparticles in a polymer matrix is a complex dynamic process which typically involves two steps. At first, Telfon AF becomes fragmented upon evaporation. Then they re-polymerize as condense on the substrate surface. In parallel, the incident vapors of metal atoms also condense on the polymer surface, which are mobile enough and can migrate along the surface or diffuse into the polymer and form clusters and particles under the cohesive force. Because of poor metal/polymer wetting, a fraction of metal atoms will desorb from the surface and return to vapor state. The ratio between the adsorbed metal atoms and the total metal atoms arriving at the substrate is called the condensation coefficient which has a small value of about 0.16 for Ag on Teflon AF,[32] indicating a relatively weak interaction between them.

Similar to the individual silver deposition, isolated individual silver nanoparticles well dispersed in the polymer matrix are the dominant features during the initial growth period for composite film thickness below *ca.* 150 Å (Figure 3d). The median particle size is about 5–9 nm depending on the metal concentration. As the nanocomposite film grows thicker, the size of embedded silver nanoparticles increases even though the metal concentration is constant. This indicates the dynamic diffusion and migration activities within the film where newly arriving metal atoms and already embedded silver clusters/particles tend to merge with each other due to their mutual affinity and weak interaction with polymers. This process might be facilitated with the thermal and kinetic energy from the vaporized metal atoms. Like the individual silver deposition, active coalescence occurs between neighboring metallic nanoparticles as the particle size increases and the interparticle spacing decreases. This leads to the growth of fracture structures with nanoclusters of more irregular shapes and broader size distributions. As a result, a significant size increase of the metal nanocrystals is observed during this stage as shown in the film of 450 Å thickness in Figure 3e, which has a median diameter of about 10–16 nm. At similar film thicknesses, larger particles and broader size distributions are observed for nanocomposites with higher metal content. In contrast to pure silver film growth, the coalescence and clustering of isolated silver nanoparticles embedded in Teflon AF occur at significantly larger film thickness. The resulted cluster size is much smaller than that in pure silver film because of the presence of polymer constituent. Thereafter, the composite films grow into a relatively stable microstructure where the dimension of nanocrystals changes very slowly (11–20 nm at a film thickness of *ca.* 1500 Å).

At film thickness above 4000 Å, a polymorphous configuration is observed, where large particle aggregates of microscale (Figure 3f) are composed of finely dispersed silver nanocrystals (inset in Figure 3f). Such superstructures are frequently observed in heterogeneous systems. In contrast to polymers, noble metals have at least two orders of magnitude higher cohesive energies.[33] Therefore metal nanoparticles have a strong tendency to come to each other. However, the presence of the extra polymer overcoat prevents them from merging into large individual grains. As a compromise, nanoscale particles are glued together by polymers into microscale aggregates.

OPTICAL ABSORPTION OF Ag/TEFLON AF NANOCOMPOSITE

The typical absorption spectrum of Ag nanoparticles of specific size is one dominant

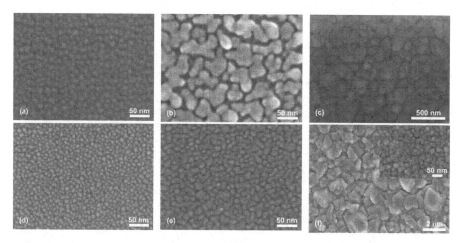

Figure 3. Representative microstructures of Ag nanoparticle films and Ag nanoparticle embedded nanocomposite films as a function of thickness. Ag films: (a) 50Å, (b) 161Å, (c) >5000Å. Ag/Teflon AF nanocomposites (73 vol% Ag loading): (d) 150Å, (e) 450Å, (f) 6650Å.

peak with a relatively narrow full width at half maximum. The peak position (resonance frequency) is strongly dependent on the size and shape of the particle as discussed above. This single resonance frequency interaction has been applied to many fields such as optical biosensors, surface-enhanced Raman scattering (SERS), and photonic crystals. For nanoscale optical biosensors, a wavelength-shift (color) response is typically measured as a result of change of inter-particle spacing or variation of the refractive index of the surface layer. A successful commercial example in this colorimetric sensing is the pregnancy test strips for home use. The results are available to naked eyes by simply comparing color difference without resorting to sophisticated instruments. These nanoparticles are also used in optical spectroscopy such as SERS and fluorescence spectroscopy. The intense localized electromagnetic fields accompanying with the surface plasmon excitation could lead to significant signal enhancement by a factor of $10^6 \sim 10^{15}$ in Raman scattering. Single-molecule detection could be enabled under the extreme enhancement condition. However, there are many other applications requiring a specific absorption profile covering a broader frequency range. Nanoparticles with one size distribution are unsuitable. By forming a group of nanoparticles with various sizes at designated content for each size, it is possible to meet this demand in theory. However, the work required to synthesize particles of each size individually and subsequently mix them together is enormous. In contrast, co-deposition of metal and polymer matrix presents a single-step mixing that can produce materials with desired absorption spectra by varying the silver loadings in the composites.

Figure 4a illustrates the absorption spectra of the Ag/Teflon AF nanocomposites. As the Ag loadings increase, the plasmon resonance frequency shifts to longer wavelength and spans a broader range covering the visible and near infrared wavelengths. With low Ag concentration, nanoparticles are isolated in majority and have a limited size distribution. Therefore the resulting absorption exhibits a relatively narrow band in the visible range, which is mainly dominated by the dipolar excitations. When the Ag concentration increases to around percolation threshold, particles and clusters of various sizes and shapes exists resulting in a fractal structure. Surface plasmon excitation could occur within the entire clusters of various sizes. Larger clusters absorb at longer wavelengths than smaller individual particles. The presence of such fractal structures therefore can strongly absorb light in a wide range of plasmon resonance frequencies.

In this study, our interests primarily focus on the exploration of their potential applications in the optoelectronics and photovoltaics. Present solar materials have an inherited weakness of partial light absorption limited by their band gap structures. For example, the most popular Si solar materials have a band gap of 1.12 eV, which results in a light absorption cut-off at about 1100 nm wavelength (Figure 4b). A large fraction of light in the near IR is not absorbed. To overcome this, tandem cells stacking multiple materials for different sub-bands in the solar spectrum are investigated. However, this complicated configuration adds significant cost with only moderate improvement. For dye-sensitized photochemical devices, multiple types of dyes with different wavelength responses have to be adopted in order to cover the solar spectrum as much as possible. Nonetheless, potential adverse cross-interference, availability as well as cost put significant limitation on this methodology.

A potential promising route to overcome the limited absorption is to harness the highly tunable optical response of metallic nanoparticles which are not limited by the band gap.[34] Figure 4b compares the absorption spectrum (violet line) with the solar radiance spectrum of global AM1.5. They represent a close match with each other. The strong light absorption and coupling from surface plasmon excitation have generated considerable interests in incorporating metal nanoparticles into Si and organic solar cells to enhance their light absorption thus overall device efficiency.[16-18,35,36] Since metal contacts are already widely used as current collectors in solar cells, the incorporation of additional plasmonic metal nanostructures can be conveniently integrated with the present fabrication setup. In more advanced design, the current collection and light coupling might be achieved in the same metal structures. Although the plasmonic silver nanoparticle can absorb the entire solar radiation, the photons with energy less than the band gaps of semiconductors are actually not utilized by them to produce photoelectrons. To overcome the limitation on light conversion, direct employment of metallic nanoparticles as photosensitizers has come into the research scope.[21] The light-induced free electron oscillation in the Ag nanoparticles generates intense electric field between the adjacent particles, which is strong enough to induce the ejection of excited electrons out of the particle surface leading to effective charge separation. The transfer of these charges to the external electrodes could be achieved via either their conduction along the percolation path of metal clusters or appropriate semiconducting medium incorporated with the nanocomposites. This possibility could open up new designs in photovoltaic devices based on the photoemission from metal nanoparticles, which we have coined as plasmon sensitized solar cells (PSSCs).

Another prominent absorption feature observed is the broadband absorption from visible to infrared regimes (red line in Figure 4a). Although the measurement range is limited by the instrument to near infrared, the present flat trend projects the absorption well beyond this regime. The numerical simulation indicates that the absorption range can extend up to ~30 μm in the far-infrared regime.[37] The present photodetectors are based on semiconductor materials which have a relatively narrow wavelength response due to the specific material band gaps. However, many applications, such as target recognition, day-night surveillance, and environmental monitoring, require multiple spectral signatures over a large wavelength range to ensure a positive identification. Owing to their unusual broad band absorption, fractal silver nanoparticles are promising candidates to develop multi-color optical sensors in a single chip. This could eliminate the expensive and cumbersome implementation of multiple sensor materials.

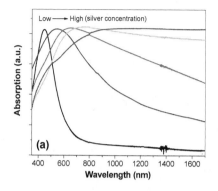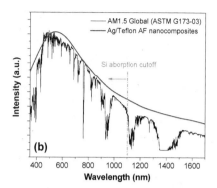

Figure 4. (a) Absorption spectra of Ag nanoparticles embedded in Teflon AF matrix as a function of Ag loadings. (b) The comparison of solar spectrum with the absorption spectrum of a Ag/Teflon AF nanocomposite of a moderate Ag loading.

ELECTRICAL PROPERTIES OF Ag NANOPARTICLE (LOADED) FILMS

The transport properties of Ag nanoparticle and its nanocomposite films are essential to understand the charge transfer after its separation by surface plasmon excitation. As their optical properties, the conductivity of Ag particle (loaded) films is closely related to the morphology which is a function of both film thickness and metal concentration.[38] Specimens for electrical measurements were deposited onto glass substrates due to their smooth and insulation nature. Two co-planar metal wire electrodes were glued to the substrate surface with water-based conductive graphite adhesives with a 10mm separation. The deposition area is 5x10 mm^2 , defined by a dielectric polytetrafluoroethylene (PTFE) mask.

Figure 5 shows the normalized electrical current in Ag films and Ag/Teflon AF nanocomposite films as a function of film thickness under the constant external potential for various metal loadings. Low potential of 1V is applied to minimize the possible stress imparted on the films under measurements. Generally, the electrical conductivity of Ag metallic films and Ag/Teflon AF nanocomposites with moderate Ag loadings (30~80%) can be divided into three stages,[39-41] which is depicted using the percolation theory where a sharp transition (percolation regime) is sandwiched between two relatively flat stages (dielectric and metallic regimes respectively). At the very thin film stage (dielectric zone), both Ag of less than 80Å thick (Zone S$_I$) and Ag/Teflon AF of less than 150Å thick (Zone C$_I$) exhibits a very low conductivity with currents on the order of a few pA. This low current is due to the isolated nature of the silver nanoparticles with small sizes and relatively large interparticle separations. Thermally activated tunneling is the dominant conduction mechanism for discontinuous and quasi-continuous metal nanoparticles below the percolation threshold since thermal activation energy is required to overcome the potential difference for electrons to transport between two particle islands. The percolation threshold is defined as the point where the first continuous electrical pathway across the film is formed.

As the film thickness increases, the interconnectivity and the formation of a fractal metallic structure (Figures 3b and e) lead to the rapid increase of electrical conductivity (Zone S_{II} of Ag and Zone C_{IIA} of nanocomposites). However, the upturn transition point occurs at much smaller film thickness for pure silver (~80Å) than for Ag/Teflon AF nanocomposites (150-200Å). For pure silver films, interparticle space contains only vacuum. While for nanocomposite films, it is filled with Teflon AF fluoropolymers which dramatically hampers the coalescence activity of the silver nanoparticles. In addition, the rate of current increase with film thickness is much faster for pure silver than Ag/Teflon AF nanocomposites. During this stage, a completely percolated metallic network is able to form in pure silver films. Accordingly, the dominant transport mechanism switches to Ohmic conduction. The percolation threshold occurs around 160–180Å in the middle of transition zone (S_{II}). In contrast, tunneling is prevalent in nanocomposites throughout the entire transition stage (Zone C_{IIA}).

In the subsequent plateau region (Zone S_{III}), silver films form a metallic continuum. The conductivity is limited by the size effect since the grain size is smaller than the electron mean free path (EMFP) (520Å for silver).[42] The electron scattering primarily occurs at the surface and grain boundary.[43,44] At thickness above 1000Å, the grain size grows larger than the EMFP. As a result, the electron–lattice scattering is dominating as in bulk silver. For composite films in the plateau Zone C_{IIB}, tunneling is still the dominant mechanism of charge transport since no metallic continuum forms yet for low or moderate metal loadings (30–80%). This is consistent with the very small increase of nanocrystal sizes previously discussed. The measured current is low on the order of nA to μA.

For nanocomposites of metal loadings higher than 90%, a second transition is present (Zone C_{IIC}). The rapidly increased current leads the nanocomposite into the flat metallic continuum regime at film thicknesses larger than 1350 Å (Zone C_{III}). Because the metal loadings are so high, a considerable amount of nanoparticles are able to interconnect with each other directly, forming extended conductive pathways. As a result, the dominant conduction mechanism switches from tunneling to metallic conduction. However, the conductivity of the resulted composites is lower than that of pure silver films because the inclusion of the dielectric polymers retains the small size of metal nanocrystals in which significant scattering occurs at grain boundaries. Compared with pure silver films with a narrow transition (Zone S_{II}), the percolation stage of nanocomposites is stretched to span a much larger film thickness (Zones C_{IIA}, C_{IIB}, and C_{IIC}) because the presence of polymers significantly slows down the interconnection activity of metal nanoparticles. A continuous metallic network is able to form only in nanocomposites with very high metal loadings.

For nanocomposites of metal loadings lower than 30%, the conductivity is relatively independent on film thickness. No rapid current change is observed. Since the metal concentration is very low, the distance between silver nanoparticles is large in comparison with their diffusion length. Moreover, the polymers filled in between further hamper their migration. As a result, the embedded metal nanoparticles are less likely become interconnected and the conductivity remains at low levels.

INTERFACIAL ELECTRONIC STRUCTURES OF SILVER/P3HT BLEND FILMS

Plasmon excited noble nanoparticles, such as gold, is shown to be able to emit photoelectrons and achieve charge separation when forming a heterojunction with appropriate semiconductors such as TiO_2 particles.[21] The charge-transfer process is largely dependent on the interfacial electronic structures between electron donors and acceptors, which is a function of their relative energy levels in valence bands and conduction bands, respectively. Thin films in the sequential layer configuration with planar extended interfaces and bulk heterojunction composites with embedded nanoparticles could exhibit dramatically different electronic properties due to the distinct interface structures. The prior one has been widely studied with ultraviolet/X-ray photoelectron spectroscopy (UPS/XPS) techniques. In contrast, the later is less studied owing to its complex nature. However, the bulk heterojunction structure is important in both fundamental science and practical applications. For example, it is widely utilized in photovoltaic devices to maximize donor-acceptor contact areas as an approach to improve the light conversion efficiency. In order to shed light in this important field, we investigate the thin film blend of P3HT containing silver nanoparticles with photoemission spectroscopy.

Figure 6 illustrates the energy diagram of Ag, P3HT and Ag/P3HT nanocomposites of various Ag content with respect to bulk silver. The vacuum level shift (Δ) of nanocomposites is variable in contrast to thin film P3HT. This provides a wide range of tunability for work function (Φ) and valence band maximum (VBM). As the polymer content increases, the work function of the blend films decreases from 3.77 to 3.37 eV, whereas the energy (ε_v^F) of the VBM relative to the Fermi level of Ag increases from 0.55 to 1.36 eV. The comparison with pure P3HT found that the corresponding parameters of pure P3HT films are not boundaries on polymer-rich side, indicating that the change of these parameters with metal or polymer content is not monotonic. The conduction band minimum (CBM) is not shown on the energy diagram since the band gap information is not available for the blend films at present. Nonetheless, they are expected to have similar variation as VBM. The very dissimilar energy levels in bulk heterojunction would have significant impact on the charge transfer process. Complimentary XPS and atomic force microscopy (AFM) investigations indicate that the co-deposition of Ag with P3HT forms complicated chemical binding states at the junction of the two constituents and composition-dependent microstructures.[28] Unlike layered structures, the interaction between Ag and S in the P3HT is much stronger for blend films due to significant larger interfaces and effective mixing of the two species. As a result, large amount of Ag-S complex exists as a "third phase" in addition to the original constituents. Therefore, the resulted nanocomposites cannot be treated as a simple physical mixture of different components. The results in Figure 6 demonstrate the

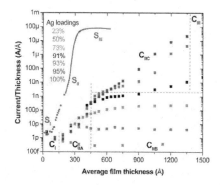

Figure 5. Normalized electrical current of Ag and Ag/Teflon AF thin films as a function of Ag loadings and film thickness.

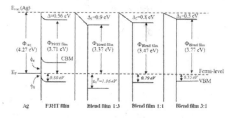

Figure 6. Schematic energy diagram of bulk Ag, P3HT film, and Ag/P3HT blend films with various Ag loadings. The composition ratio is metal to polymer in volume. See details in the main text. (Reused from Ref. 28 with permission, Copyright 2009, American Chemical Society)

feasibility to tune the energy levels by varying the composition of the blend film. This adds complexity but also versatility in the design of devices in tailoring the charge injection barriers.

CONCLUSION

Thin film nanocomposites containing silver nanoparticles embedded in dielectric and semiconducting polymer matrix were fabricated using the vapor phase co-deposition technique. Optical, electrical, and interfacial electronic properties are evaluated, which are strongly dependent on the microstructures of metal particles. As the silver loading increases from low to high, a variety of tailorable absorption profiles are achievable including a solar matching absorption spectrum and an unusual broadband absorption from visible to infrared regimes. This is of particular interest for developing highly efficient solar cells and multi-spectral photodetectors, respectively. Like metal films, the electrical conductivity of silver/polymer composites can be divided into three zones with distinct microstructures: dielectric (isolated particle islands), transition (percolated metallic network), and metallic (metallic continuum). However, the transition is significantly stretched in comparison with pure silver films as a result of the presence of polymer inclusion. Nanocomposites in bulk heterojunction structures exhibit very different interfacial electronic structures from layered configurations due to the vast interfaces present which promote chemical interactions resulting in considerable amount of extra third phases. This provides a viable route to tailor the electronic properties such as work function and barrier height by varying the composition of the blend film, which could possibly fulfill the potential of metal nanoparticles as photosensitizers in photoelectron generation in addition to light concentration and dramatically improve the efficiency of light harvesting devices.

ACKNOWLEDGEMENTS

The authors gratefully acknowledge the financial support from the ARO Grant W911NF-06-1-0295 and the ONR Grant N00014-03-1-0247. The authors would like to thank Elsevier for the permission to reuse Figures 3f and Figure 5 from Ref. 38 under the journal authors' rights.

REFERENCES
[1] Y. N. Xia and N. J. Halas, Shape-Controlled Synthesis and Surface Plasmonic Properties of Metallic Nanostructures, *MRS Bull.*, **30**, 338-44 (2005).
[2] M. Faraday, The Bakerian Lecture: Experimental Relations of Gold (and Other Metals) to Light, *Philos. Trans. R. Soc. London*, **147**, 145-81 (1857).
[3] G. Mie, Contributions to the Optics of Turbid Media, Particularly of Colloidal Metal Solutions, *Ann. Phys. (Leipzig)*, **330**, 377-445 (1908).
[4] U. Kreibig and L. Genzel, Optical-Absorption of Small Metallic Particles, *Surf. Sci.*, **156**, 678-700 (1985).
[5] D. M. Kuncicky, B. G. Prevo, and O. D. Velev, Controlled Assembly of SERS Substrates Templated by Colloidal Crystal Films, *J. Mater. Chem.*, **16**, 1207-11 (2006).
[6] M. Moskovits, Surface-Enhanced Raman Spectroscopy: A Brief Retrospective, *J. Raman Spectrosc.*, **36**, 485-96 (2005).
[7] A. Biswas, O. C. Aktas, U. Schurmann, U. Saeed, V. Zaporojtchenko, F. Faupel, and T. Strunskus, Tunable Multiple Plasmon Resonance Wavelengths Response from Multicomponent Polymer-Metal Nanocomposite Systems, *Appl. Phys. Lett.*, **84**, 2655-57 (2004).

[8]Y. Dirix, C. Bastiaansen, W. Caseri, and P. Smith, Oriented Pearl-Necklace Arrays of Metallic Nanoparticles in Polymers: A New Route toward Polarization-Dependent Color Filters, *Adv. Mater.*, **11**, 223-27 (1999).

[9]M. Quinten, The Color of Finely Dispersed Nanoparticles, *Appl. Phys. B: Lasers Opt.*, **73**, 317-26 (2001).

[10]G. I. Stegeman and E. M. Wright, All-Optical Wave-Guide Switching, *Opt. Quantum Electron.*, **22**, 95-122 (1990).

[11]A. Biswas, H. Eilers, F. Hidden, O. C. Aktas, and C. V. S. Kiran, Large Broadband Visible to Infrared Plasmonic Absorption from Ag Nanoparticles with a Fractal Structure Embedded in a Teflon Af (R) Matrix, *Appl. Phys. Lett.*, **88**, 013103 (2006).

[12]C. H. Shek, G. M. Lin, J. K. L. Lai, and J. L. Li, Fractal Structure and Optical Properties of Semicontinuous Silver Films, *Thin Solid Films*, **300**, 1-5 (1997).

[13]V. M. Shalaev, *Optical Properties of Nanostructured Random Media*, Vol. 82 (Springer, Berlin, 2002).

[14]Y. Yagil and G. Deutscher, Transmittance of Thin Metal-Films near the Percolation-Threshold, *Thin Solid Films*, **152**, 465-71 (1987).

[15]Y. Yagil, P. Gadenne, C. Julien, and G. Deutscher, Optical-Properties of Thin Semicontinuous Gold-Films over a Wavelength Range of 2.5 to 500 μm, *Phys. Rev. B*, **46**, 2503-11 (1992).

[16]M. Westphalen, U. Kreibig, J. Rostalski, H. Luth, and D. Meissner, Metal Cluster Enhanced Organic Solar Cells, *Sol. Energy Mater. Sol. Cells*, **61**, 97-105 (2000).

[17]S. Pillai, K. R. Catchpole, T. Trupke, and M. A. Green, Surface Plasmon Enhanced Silicon Solar Cells, *J. Appl. Phys.*, **101**, 093105 (2007).

[18]H. A. Atwater and A. Polman, Plasmonics for Improved Photovoltaic Devices, *Nat. Mater.*, **9**, 205-13 (2010).

[19]W. L. Barnes, Electromagnetic Crystals for Surface Plasmon Polaritons and the Extraction of Light from Emissive Devices, *J. Lightwave Technol.*, **17**, 2170-82 (1999).

[20]K. R. Catchpole and S. Pillai, Surface Plasmons for Enhanced Silicon Light-Emitting Diodes and Solar Cells, *J. Lumin.*, **121**, 315-18 (2006).

[21]Y. Tian and T. Tatsuma, Mechanisms and Applications of Plasmon-Induced Charge Separation at TiO_2 Films Loaded with Gold Nanoparticles, *J. Am. Chem. Soc.*, **127**, 7632-37 (2005).

[22]S. Link and M. A. El-Sayed, Spectral Properties and Relaxation Dynamics of Surface Plasmon Electronic Oscillations in Gold and Silver Nanodots and Nanorods, *J. Phys. Chem. B*, **103**, 8410-26 (1999).

[23]A. Moores and F. Goettmann, The Plasmon Band in Noble Metal Nanoparticles: An Introduction to Theory and Applications, *New J. Chem.*, **30**, 1121-32 (2006).

[24]P. Mulvaney, Surface Plasmon Spectroscopy of Nanosized Metal Particles, *Langmuir*, **12**, 788-800 (1996).

[25]S. Padovani, C. Sada, P. Mazzoldi, B. Brunetti, I. Borgia, A. Sgamellotti, A. Giulivi, F. D'Acapito, and G. Battaglin, Copper in Glazes of Renaissance Luster Pottery: Nanoparticles, Ions, and Local Environment, *J. Appl. Phys.*, **93**, 10058-63 (2003).

[26]T. C. Nason, J. A. Moore, and T. M. Lu, Deposition of Amorphous Fluoropolymer Thin-Films by Thermolysis of Teflon Amorphous Fluoropolymer, *Appl. Phys. Lett.*, **60**, 1866-68 (1992).

[27]G. B. Blanchet, Deposition of Amorphous Fluoropolymers Thin-Films by Laser Ablation, *Appl. Phys. Lett.*, **62**, 479-81 (1993).

[28]L. Scudiero, H. Wei, and H. Eilers, Photoemission Spectroscopy and Atomic Force Microscopy Investigation of Vapor-Phase Codeposited Silver/Poly(3-Hexylthiophene) Composites, *ACS Applied Materials & Interfaces*, **1**, 2721-28 (2009).

[29]H. Wei, L. Scudiero, and H. Eilers, Infrared and Photoelectron Spectroscopy Study of Vapor Phase Deposited Poly (3-Hexylthiophene), *Appl. Surf. Sci.*, **255**, 8593-97 (2009).

[30]H. Wei and H. Eilers, From Silver Nanoparticles to Thin Films: Evolution of Microstructure and Electrical Conduction on Glass Substrates, *J. Phys. Chem. Solids*, **70**, 459-65 (2009).

[31]R. D. Fedorovich, A. G. Naumovets, and P. M. Tomchuk, Electron and Light Emission from Island Metal Films and Generation of Hot Electrons in Nanoparticles, *Phys. Rep.*, **328**, 73-179 (2000).

[32]H. Takele, H. Greve, C. Pochstein, V. Zaporojtchenko, and F. Faupel, Plasmonic Properties of Ag Nanoclusters in Various Polymer Matrices, *Nanotechnology*, **17**, 3499-505 (2006).

[33]K. S. Giesfeldt, R. M. Connatser, M. A. De Jesus, P. Dutta, and M. J. Sepaniak, Gold-Polymer Nanocomposites: Studies of Their Optical Properties and Their Potential as SERS Substrates, *J. Raman Spectrosc.*, **36**, 1134-42 (2005).

[34]H. Eilers, A. Biswas, T. D. Pounds, M. G. Norton, and M. Elbahri, Teflon AF/Ag Nanocomposites with Tailored Optical Properties, *J. Mater. Res.*, **21**, 2168-71 (2006).

[35]B. P. Rand, P. Peumans, and S. R. Forrest, Long-Range Absorption Enhancement in Organic Tandem Thin-Film Solar Cells Containing Silver Nanoclusters, *J. Appl. Phys.*, **96**, 7519-26 (2004).

[36]R. F. Service, Solar Energy: Can the Upstarts Top Silicon?, *Science*, **319**, 718-20 (2008).

[37]D. A. Genov, A. K. Sarychev, and V. M. Shalaev, Metal-Dielectric Composite Filters with Controlled Spectral Windows of Transparency, *Journal of Nonlinear Optical Physics & Materials*, **12**, 419-40 (2003).

[38]H. Wei and H. Eilers, Electrical Conductivity of Thin-Film Composites Containing Silver Nanoparticles Embedded in a Dielectric Fluoropolymer Matrix, *Thin Solid Films*, **517**, 575-81 (2008).

[39]K. Seal, M. A. Nelson, Z. C. Ying, D. A. Genov, A. K. Sarychev, and V. M. Shalaev, Growth, Morphology, and Optical and Electrical Properties of Semicontinuous Metallic Films, *Phys. Rev. B*, **67**, 035318 (2003).

[40]C. A. Neugebauer and M. B. Webb, Electrical Conduction Mechanism in Ultrathin, Evaporated Metal Films, *J. Appl. Phys.*, **33**, 74-82 (1962).

[41]A. Kiesow, J. E. Morris, C. Radehaus, and A. Heilmann, Switching Behavior of Plasma Polymer Films Containing Silver Nanoparticles, *J. Appl. Phys.*, **94**, 6988-90 (2003).

[42]W. Zhang, S. H. Brongersma, O. Richard, B. Brijs, R. Palmans, L. Froyen, and K. Maex, Influence of the Electron Mean Free Path on the Resistivity of Thin Metal Films, *Microelectron. Eng.*, **76**, 146-52 (2004).

[43]W. Wu, S. H. Brongersma, M. Van Hove, and K. Maex, Influence of Surface and Grain-Boundary Scattering on the Resistivity of Copper in Reduced Dimensions, *Appl. Phys. Lett.*, **84**, 2838-40 (2004).

[44]P. M. T. M. van Attekum, P. H. Woerlee, G. C. Verkade, and A. A. M. Hoeben, Influence of Grain-Boundaries and Surface Debye Temperature on the Electrical-Resistance of Thin Gold-Films, *Phys. Rev. B*, **29**, 645-50 (1984).

Nanolaminated
Ternary Carbides

TRIBOFILM FORMATION USING Ti$_2$AlC MATERIAL

P. Kar, S. Kundu, M. Radovic, and H. Liang
Department of Mechanical Engineering, Texas A&M University
College Station, TX 77843

ABSTRACT
Formation of tribofilm is important for anti-wear behavior. In this research, we used a new approach to reduce wear. Specifically, we applied Ti$_2$AlC materials as additives during wear of medium carbon steel. Ti$_2$AlC like the rest of MAX phases is nano-layered materials with low coefficient of friction, low thermal expansion, good thermal and electrical conductivities, and is resistant to chemical attack. Instead of wear, it was found that there was a Ti$_2$AlC tribofilm formed on the steel surface protecting material loss. Detailed characterization was conducted over the film properties. In this paper, we discuss mechanisms of Ti$_2$AlC tribofilm formation and wear that are important for tribological applications.

INTRODUCTION
More than 50 known MAX phases - a new family of nanolayered thermodynamically stable ternary carbides and nitrides with the general formula $M_{n+1}AX_n$ - MAX for short where M is an early transition metal, A is an A-group element (mostly 3A and IV A) and X is C and/or N and n = 1 to 3 - have hexagonal crystal structure and unique set of properties - some of which are typical of ceramics, while others are more typical of metals.[1-3] Like most of ceramics, they have low coefficient of thermal expansion that ranges from ≈ 7.5 x 10^{-6} K^{-1} to ≈ 13 x 10^{-6} K^{-1} and are stable up to at least 1600°C (Ti$_3$SiC$_2$ is stable up to 2200 °C). Despite being elastically quite stiff[3-6] these solids are relatively soft (Vickers hardness 2-5 GPa) and most readily machinable with regular high-speed tool steel with no lubrication or cooling required.[1-3] Due to their layered atomic structure and the presence of active slip systems, the MAX phases possess unique mechanical properties that are atypical for ceramics.[1-3] By now, the exceptional thermal shock resistant and damage tolerance of the MAX phases has been well established.[3] The fracture toughness of Ti$_3$SiC$_2$ varies from 8-16 MPam$^{1/2}$ depending on crack length.[7] The latter value is one of the highest ever reported for a single-phase, monolithic ceramic. Recently, it was found that MAX phases, in particular Ti$_3$SiC$_2$ although elastically stiff can dissipate up to 25% of the mechanical energy when loaded in compression and tension due to hysteretic, non-linear behavior even at room temperatures.[8,9] The practical implication of this atypical property, cannot be overemphasized since the material's ability to absorb, and thus not transmit vibration, or noise, is related to its damping capacity.[3] At high temperatures, above $\approx$1000 °C, the MAX phases deform plastically even in tension, with strains to failure that exceed 25%.[9,10] However, despite being relatively soft and ductile at elevated temperatures, the MAX phases are quite creep resistant. The creep studies of the Ti$_3$SiC$_2$ in 1000-1200 °C suggest that their creep and oxidation resistance is comparable, or even better than those of the most high temperature metallic alloys.[9-13] By now, it is also well established that the MAX phases can also form a large number of solid solutions, by substitutions on the M and/or A and/or X sites. Here again the importance of this fact cannot be overemphasized since this will allow us to tailor the properties

of the MAX phases by making solid solutions, in a fashion that is more akin to metallic alloys than traditional ceramics.[1]

Among all those properties, low coefficient of friction, low wear and high resistance to oxidation and corrosion, make MAX phases good candidates for formation of protective tribofilms on metallic substrates. The self-lubricant properties of Ti$_3$SiC$_2$ were firstly observed by Barsoum et al.[14] and they were attributed to the layered structure with weak bonding between layers, i.e. the structure similar to that of graphite. The friction coefficients of the Ti$_3$SiC$_2$ were measured by lateral force microscopy to be 2-5 x 10^{-3} for the basal planes - the one of the lowest kinetic friction coefficients ever reported.[15-16] This low coefficient of friction remained that low after up to 6 months exposure to the atmosphere. Unfortunately, significantly higher coefficients of friction were measured for polycrystalline Ti$_3$SiC$_2$ samples using discs-on-pin sliding wear tests against 440C steel pin and in diamond belt abrasion tests.[17] It was found that polycrystalline Ti$_3$SiC$_2$ undergoes an initial transition stage where coefficient of friction, μ, increases linearly to 0.15 - 0.45 after which it rises to steady state values of 0.83. The average sliding wear rate measured were 4.25x10^{-3} and 1.34x10^{-3} mm^3/Nm for fine and coarse grained microstructure, respectively. In the diamond belt abrasion tests, the average wear rates were much higher, i.e. 6.14x10^{-2} and 3.96 x10^{-2} mm^3/Nm for fine and coarse grained microstructure, respectively.[17] The wear occurs by grain prefracture, grain pull-outs, delamination and grain deformation. In the later studies, two successive friction regimes have been also identified for both fine and course grained Ti$_3$SiC$_2$ in tribological tests against steel and Si$_3$N$_4$ balls.[18] In this study formation of the tribofilm was reported for both counterparts. Most recently, tribological behavior of series of MAX phases (Ti$_2$AlC, Cr$_2$AlC, Ta$_2$AlC, Ti$_3$SiC$_2$, Ti$_2$AlN, Ti4AlN$_3$, Cr$_2$GeC, Cr$_2$GaC, Nb$_2$SnC and Ti$_2$SnC) was tested against Ni-based superalloys at 25 and 550 °C.[19] It was suggested that the relatively high wear rates of polycrystalline MAX phases at room temperature occurs due to third body abrasion. However, ultra low wear rates at high temperatures were attributed to the lubricious tribooxide films, comprised mainly of oxides of the superalloy elements, viz. Ni, Fe and Cr.[19] In the dry sliding tribologycal tests against alumina at 550 °C, formation of the tribofilm comprised of amorphous oxides of the M and A elements and, in some cases, unoxidized grains of the corresponding MAX phases, was observed on the contact surfaces, for several different MAX phases.[20] Most recently, good tribological performance of two new composite materials, consisting of MAX phases, (Ta$_2$AlC or Cr$_2$AlC) and 20 vol.% Ag, were investigated against a Ni-based superalloy and alumina and reported results suggest that those composites could be promising materials for various high temperature tribological applications.[20]

Another attractive property of MAX phases, that is of crucial importance for protection of metallic surfaces, is their good corrosion and oxidation resistance. Good corrosion resistance of the several MAX phases in acids and alkalis such as HCl, H$_2$SO$_4$, NaOH and KOH at room temperatures were demonstrated in several papers.[21-24] In addition to good corrosion resistance, some of MAX phases such as Ti$_3$SiC$_2$ [25] and Ti$_2$AlC [26,27] have high oxidation resistance at high temperatures up to 1300 °C. For example, Ti$_2$AlC forms a protective, predominately Al$_2$O$_3$ scale at elevated temperatures in oxidizing environment that does not spall off, or separate from, Ti$_2$AlC during severe thermal cycling.[27]

In this paper we report on the formation of tribofilm as a result of tribochemical reaction between Ti$_2$AlC and a medium carbon steel under sliding conditions.

EXPERIMENTAL PROCEDURE

Lubricants

Two types of oils were used as carriers, namely the mineral oil and castor oil. The mineral oil is non-polar that has been used previously as base oil for tribological applications.[28] This oil is made mostly of hydrocarbons.[29] The castor oil has been used as a base oil for similar reasons and it is polar containing a hydroxyl group in its carbon chain.[30] TiAl$_2$C (supplier 3-one-2, LLC, NJ) is used as the additives and its properties are being tested on ferrous-based alloys.

Materials

A medium carbon steel disk was slid against a bearing steel ball for tribology tests in the pin-on-disk configuration in a reciprocal motion. Before experiments, disks were cleaned with acetone and singed with water after touch polishing with 800-1200 grit sand papers. The pin was an ES2100 steel ball with a nominal radius of 6 mm. To study the effects of third body on wear and tribofilm formation, 50 %wt of Ti$_2$AlC powder of 325 mesh was in the lubricant formulation as abrasives and potential bonding agent for tribofilms. Ti$_2$AlC powder was predominately single phase with 2-3% impurities manly Ti$_3$AlC$_2$. During sliding, when added to the lubricant, the Ti$_2$AlC powder could either be deformed or undergo tribochemical reactions. It is known that transition metals have an affinity for electron rich organic molecules from lubricating oil that can promotes formation of tribofilms.[31]

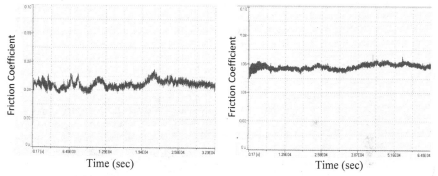

Fig. 1. Friction coefficient as a function of test time. Fig. 1a on the left is obtained from a test in the castor oil; Fig. 1b on the right is from the mineral oil. The y-axis is coefficient of friction and x-axis is time in seconds.

Test Apparatus

A CSM pin-on-disk tribometer was used for this of the interface between the test sample (medium carbon steel) and its static partner (pin), ES2100 steel ball. A reciprocal motion was used during experiments. A series of sliding experiments were performed using different lubricant formulations. A normal load of 10 N was applied to the medium carbon steel surface. The corresponding Hertzian contact pressure was around 4 GPa. The sliding speed was 5 cm/sec excluding at the two end points of the wear track. A boundary lubrication regime was maintained so that a tribofilm formation is made possible. The friction coefficient, μ, was determined by dividing the tangential force F_t by the applied normal force F_N, i.e., $\mu = F_t/F_N$. The test duration was set to 4 hours in order to obtain a sufficiently thick tribofilm. The total length of the wear track was 6 mm.

Characterization

A JEOL JSM 6400 Scanning Electron Microscope (SEM) was used for wear mode study. A tungsten filament was used and the operating voltage was 15 KV. Chemical characterization of the tribofilm was done using Energy Dispersive X-ray Spectroscopy (EDX).

RESULTS AND DISCUSSIONS

The friction coefficient as a function of time obtained during triboexperiments is shown in Figure 1. The Figure 1a shows the result obtained in the castor oil while Figure 1b from the mineral oil. The X-axis is time (seconds) and Y-axis the friction coefficient. The friction coefficient, μ was ranged from 0 to 0.1. All test conditions are the same except base oil used. As it can be seen in Figure 1, the castor oil generates a lower friction than the mineral oil. As mentioned earlier, the castor oil molecules are polar while the minerals are non-polar. The majority of their molecular structures are illustrated in the following.

Castor oil:

Mineral oil:
$$\left(- CH_2 - CH_2 - CH_2 - \right)_n$$

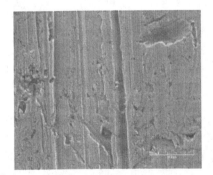 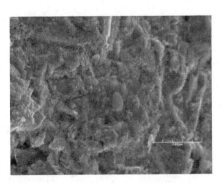

Fig. 2.: The formation of a tribofilim after tests in mineral oil. Fig. 2a on the left is the morphology of the film. The Fig. 2b on the right shows the microstructures of the film. The scale bar on the left is 20 μm and right 7 μm.

A low friction coefficient indicates that the attachment of castor molecules to the sample surface promoted lubrication. While lack of attachment of mineral oil molecules to the either steel surface or the Ti₂AlC power means that they are rather freely rubbing against each other. During sliding, the direct physical contact of steel surfaces and Ti₂AlC power enabled the joint of their surfaces, hence the formation of tribofilms. Interestingly, the power particles of Ti₂AlC did not work as abrasives thus no wear was generated.

After triboexperiments, two samples tested in two different oils exhibited different types of wear track. The wear track obtained in castor oil shows scratches while that formed in the mineral oil has a tribofilm on its surface, instead of wear. The characterization of the wear track from the mineral oil was conducted using the SEM, as shown in Figure 2. The Fig. 2a on the left shows a continued film covering the surface of the steel sample. The visible grooves indicate the movement of film under sliding. The magnification of 4,500X in Fig. 2b, it shows the elongated grain like structures typical for Ti₂AlC phase.

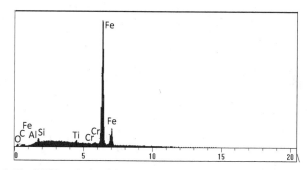

Fig. 3. The EDX analysis of the tested steel surface. The surface was covered by a tribofilm.

Table 1, Elements shown in the EDX spectra.

Element	Si	Ti	Fe	Al	C	Cr	Total
Wt%	0.93	0.74	77.91	0.12	20.07	0.24	100.00

The EDX analysis was carried out on the surface of the steel sample tested in mineral oil to determine chemical composition of the tribofilm. The results are shown in the Figure 3 and Table 1. As it can be seen, the major elements are Fe and Si that originate from the low carbon steel. However, detected Ti, Al, and C were from the Ti₂AlC phase and lubricant. Apparently, the tribofilm formed due to the stir and mechanical mixing resulted in the binding of materials at the interface. Currently, the properties of the tribofilm and binding mechanisms are under investigation. The characteristic microstructure of the Ti₂AlC powder shown in Figure 2a provided the evidence that the tribofilm was made of the Ti₂AlC material.

CONCLUSIONS

We conducted tribological experiments to study the interactions between Ti₂AlC and steel surfaces under sliding in two different lubricants, mineral and castrol oils. In the case of the castrol oil, abrasive wear was observed with a relatively low friction. In the case of mineral oil, a tribofilm was formed, instead of wear. The film was observed on the worn surface that is smooth and uniform. The tribofilm was apparently made of the Ti$_2$AlC phase. The mineral oil was found to work as a vehicle to enable the physical contact needed for the tribofilm formation.

Acknowledgement
This work was in part sponsored by the National Science Foundation (0535578).

REFERENCES
1. M.W. Barsoum, The M$_{N+1}$AX$_N$ Phases: New Class of Solids; Thermodynamically Stable Nanolaminates, *Prog. Solid State Chem.* **28**, 201-281 (2000).
2. M.W. Barsoum, Physical properties of the MAX phases, Encyclopedia of Materials Science and Technology, Eds. K. H. J. Buschow, R. W. Cahn, M. C. Flemings, E.J. Kramer, S. Mahajan and P. Veyssiere, Elsevier Science, Amsterdam, (2004).
3. M.W. Barsoum and M.Radovic, Mechanical Properties of the MAX Phases Encyclopedia of Materials Science and Technology, Eds. K. H. J. Buschow, R. W. Cahn, M. C. Flemings, E. J. Kramer, S. Mahajan and P. Veyssiere, Elsevier Science, Amsterdam, (2004).
4. M. Radovic, M.W. Barsoum, A. Ganguly, T. Zhen, P. Finkel, S. R. Kalidindi and E. Lara-Curzio, On the Elastic Properties and Mechanical Damping of Ti$_3$SiC$_2$, Ti$_3$GeC$_2$, Ti$_3$Si$_{0.5}$Al$_{0.5}$C$_2$ and Ti$_2$AlC, in the 300-1573 K Temperature Range, *Acta. Mat.* **54**, 2757-2767 (2005).
5. M. Radovic, A. Ganguly and M. W. Barsoum, Elastic properties and phonon conductivities of Ti$_3$Al(C$_{0.5}$,N$_{0.5}$)$_2$ and Ti$_2$Al(C$_{0.5}$,N$_{0.5}$) solid solutions, J. Mater. Res. 23 (2008) 1517-1521.
6. M. Barsoum, M. Radovic, T. Zhen, P. Finkel and S. Kalidindi, Dynamic Elastic Hysteretic Solids and Dislocations, *Phys. Rev. Lett.* **94**, 085501-1(2005).
7. C. J. Gilbert, D. R. Bloyer, M. W. Barsoum, T. El-Raghy, A. P. Tomsia and R. O. Ritchie, Fatigue-crack growth and fracture properties of coarse and fine-grained Ti$_3$SiC$_2$, *Scripta Mat.* **128**, 125 (2000).
8. M.W. Barsoum, T. Zhen, S. Kalidindi, M. Radovic and A. Murugaiah, Fully Reversible Plasticity Up to 1 GPa in Ti$_3$SiC$_2$, *Nature Materials* **2**, 107-111 (2003).
9. M. Radovic, M.W. Barsoum, T. El-Raghy, J. Seidensticker, S. Wiederhorn, Tensile Properties of Ti$_3$SiC$_2$ in the 25-1200°C Temperature Range, *Acta Mat.* **48**, 453-459 (2000).
10. M. Radovic, M.W. Barsoum, T. El-Raghy, S. Wiederhorn and W. Luecke, Effect of Temperature, Strain Rate and Grain Size on the Mechanical Response of Ti$_3$SiC$_2$ in Tension, *Acta Mat.* **50**, 1297-1306 (2002).
11. M. Radovic, M.W. Barsoum, T. El-Raghy, S. Wiederhorn, Tensile Creep of Fine Grained Ti$_3$SiC$_2$ in the 1000-1200°C Temperature Range, *Acta Mat.* **49**, 4103-4112, (2001).
12. M. Radovic, M.W. Barsoum, T. El-Raghy, S. Wiederhorn, Tensile Creep of Coarse Grained Ti$_3$SiC$_2$ in the 1000-1200 °C Temperature Range, *J. of Alloys Compds.* **361**, 299-312 (2003).
13. T. Zhen, M. W. Barsoum and S. R. Kalidindi, M. Radovic, Z. M. Sun, T. El-Raghy, Compressive Creep of Fine and Coarse-Grained Ti$_3$SiC$_2$ in Air in the 1100 to 1300 °C Temperature Range, *Acta. Mater.* **53**, 4963-4973, (2005).
14. M.W. Barsoum and T. El-Raghy, Synthesis and characterization of a remarkable ceramic: Ti$_3$SiC$_2$, *J. Amer. Ceram. Soc.* **79**, 1953 (1996).
15. S. Myhra , J.W.B.Summers, E.H. Kisi, Ti$_3$SiC$_2$ - A layered ceramic exhibiting ultra-low friction, *Materials Letters.* **39**, 6-11, (1999).

16. A. Crossley, E.H. Kisi, J. W. B. Summers JWB, S. Myhra , Ultra-low friction for a layered carbide-derived ceramic, Ti$_3$SiC$_2$, investigated by lateral force microscopy (LFM), *J. of Phys.*, **D32** 632-638 (1999).

17. T. El-Raghy, P. Blau, M.W. Barsoum, Effect of grain size on friction and wear behavior of Ti$_3$SiC$_2$, *Wear* **42**, 761 (2000).

18. A. Soucheta, J. Fontainea, M. Belina, T. Le Mognea, J.-L. Loubeta, M.W. Barsoum, Tribological duality of Ti$_3$SiC$_2$, *Tribology Letters* **18**, 341(2005).

19. S. Gupta, D. Filimonov, V. Zaitsev, T. Palanisamy, M.W. Barsoum, Ambient and 550 °C tribological behavior of select MAX phases against Ni-based superalloys, *Wear* **264**, 270–278, (2008).

20. S. Gupta, D. Filimonov, T. Palanisamy, M.W. Barsouma, Tribological behavior of select MAX phases against Al$_2$O$_3$ at elevated temperatures, *Wear* **265**, 560–565 (2008).

21. V.D. Jovica, M. W. Barsoum, Corrosion Behavior and Passive Film Characteristics Formed on Ti, Ti$_3$SiC$_2$, and Ti$_4$AlN$_3$ in H$_2$SO$_4$ and HCl , *J. Electrochem. Soc.* **151**, 4274-4282 (2004).

22. J. Travaglini, M.W. Barsoum, V. Jovic, T. El-Raghy, The corrosion behavior of TI$_3$SiC$_2$ in common acids and dilute NaOH, *Corrosion Sci.* **45**, 1313 -1327 (2003).

23. V.D. Jovic, M.W. Barsoum, B.M. Jovic, A. Ganguly and T. El-Raghy, Corrosion behavior of Ti$_3$GeC$_2$ and Ti$_2$AlN in 1M NaOH, *J. Electrochem. Soc.* **153**, B238-B243 (2006).

24. V.D. Jovic, B.M. Jovic, S. Gupta, T. El-Raghy, M.W. Barsoum, Corrosion behavior of select MAX phases in NaOH, HCl and H$_2$SO$_4$, *Corrosion Sci.* **48**, 4274-4282 (2006).

25. M.W. Barsoum, L.H. Ho-Duc, M. Radovic and T. El-Raghy, Long Time oxidation Study of Ti$_3$SiC$_2$, Ti$_3$SiC$_2$/SiC and Ti$_3$SiC$_2$/TiC Composites in Air, *J Electrochem. Soc.* **150**, 166-175 (2003).

26. J.W. Byeon, J. Liu, M. Hopkins, W. Fischer, K.B. Park, M.P. Brady, M. Radovic, T. El-Raghy, Y.H. Sohn, Microstructure and Residual Stress of Alumina Scale Formed on Ti2AlC at High Temperature in Air, *Oxidation of Metals.* **68**, (2007) 97-111.

27. M. Sundberg, G. Malmqvist, A. Magnusson, T. El-Raghy, Alumina forming high temperature silicides and carbides, *Ceramics International.* **30**, 1899-1904(2004).

28. P. Kar, P. Asthana, and H. Liang, Formation and Characterization of Tribofilms, *J. Tribo.* **130**, 04201-042306 (2008).

29. J.E. Anderson, B.R. Kim, S.A. Mueller, T.V. Lofton, Composition and analysis of mineral oils and other organic compounds in metalworking and hydraulic fluids, *Critical Reviews in Environmental Science and Technology* **33**, 73-109 (2003).

30. J. Pokorný, J. Hladík J, I. Zeman, Composition of Castor Oil, *Pharmazie* **23**, 332-333 (1968).

31. F. Muktepavela, G. Bakradze, E. Tamanis, S. Stolyarova, and N. Zaporina, Influence of Mechanoactivation on the Adhesion and Mechanical Properties of Metal/oxide Interfaces, *Phys. Stat. Sol.* **2**, 339 – 342 (2005).

Author Index

Author Index

Vendrame, S. C., 69

Walz, J. Y., 57
Wang, H., 143
Wang, X., 163

Wei, H., 171
Wen, J. Z., 115
Wu, N., 163

Yan, Y., 143